Resurgence of Nuclear Power

Nandakumar Janardhanan
Girijesh Pant · Ravi B. Grover
Editors

Resurgence of Nuclear Power

Challenges and Opportunities for Asia

Springer

Editors
Nandakumar Janardhanan
Energy Studies Programme
Jawaharlal Nehru University
New Delhi, Delhi
India

Ravi B. Grover
Homi Bhabha National Institute
Mumbai, Maharashtra
India

Girijesh Pant
Energy Studies Programme
Jawaharlal Nehru University
New Delhi, Delhi
India

ISBN 978-981-10-5028-2 ISBN 978-981-10-5029-9 (eBook)
https://doi.org/10.1007/978-981-10-5029-9

Library of Congress Control Number: 2017943099

Printed on acid-free paper

This Springer imprint is published by Springer Nature
The registered company is Springer Nature Singapore Pte Ltd.
The registered company address is: 152 Beach Road, #21-01/04 Gateway East, Singapore 189721, Singapore

Preface

Fossil fuels remain the largest source of energy-related emissions. This has not only led the countries across the world to explore ways to minimise dependence on them, but has also given thrust to the development of cleaner energy sources such as renewable and nuclear. Following commitments made by various countries in Paris in December 2015, demand for low-carbon energy sources is increasing and energy experts as well as the civil society recognise the need to move away from fossil fuels. In the low-carbon energy landscape, renewable sources and nuclear power are emerging as the dominant players.

In addition to being a low-carbon source, nuclear power offers several advantages. It provides baseload power at competitive tariff, has low external costs, has a high value of EROI (Energy returned on invested) and it is possible to have plant size varying from about 200 MW to about 1500 MW. Low land requirements and high energy density of nuclear fuels are its still further advantages. As electricity demand increases, developing countries are looking at increasing the share of nuclear in their energy mix. The main challenges faced by nuclear energy are high capital cost, which must be provided for upfront and public perception.

Solar and wind energy sources are the major renewable sources and their cost of generation is becoming comparable to fossil fuel alternatives. However, they face multiple challenges. The first challenge arises from the intermittency resulting in steep cost for integrating renewables in the grid commonly referred to as system integration costs. The second challenge arises from large land requirements and the next challenge arises from the ecological concerns related to the impact on flora and fauna. Though the cost of solar and wind facilities is continuously falling, lack of efficient and cost-effective energy storage technology is a major handicap for their large-scale expansion.

Coming back to nuclear, the Asian region is today home to the largest number of nuclear power plants under construction. According to World Nuclear Association, in East Asia through South Asia, there are 128 operating nuclear power reactors, 40 under construction and firm plans to build a further 90. While the potential for further expansion of nuclear energy sector in this region remains huge, it is also important to understand various challenges that can critically influence this sector.

This book is an attempt jointly made by eminent scientists, academicians and policy researchers to explore the nuances of the resurgence of nuclear energy in the Asian region and the underlying opportunities and challenges that the sector faces in its growth trajectory. The idea for coming up with this book took shape during the first nuclear energy seminar conducted by the Energy Studies Program of Jawaharlal Nehru University in April 2016, where some of the contributing authors participated and presented various aspects related to nuclear energy sector in the Asian region. After several rounds of discussion among the participants, it was decided to bring out a book which can project ideas and thoughts relevant to the nuclear energy sector in Asia.

The book has been divided into five sections and addresses a wide range of issues. The first part looks at the debates surrounding resurgence of nuclear power in the Asian region; second part focuses on the public perception and explores the key debates about the role of public perception in nuclear sector's growth in this region; the third part in this book examines the nuclear knowledge management, the fourth part details the legal aspects with specific cases from South Asia and the fifth part is dedicated to examining the economics of nuclear power.

The first chapter focuses on the resurgence of nuclear power in India. Various available energy supply options are examined in the backdrop of economic as well as resource sustainability. The author argues that nuclear energy will play a significant role in the low-carbon energy mix in India. The chapter links economic as well as human development aspects and estimates the significance of nuclear power from the perspective of environmental sustainability and affordability. The chapter also outlines nuclear cooperation agreement entered by India with countries having advanced nuclear technology as well as countries exporting uranium. Further the chapter examines the programmes of the Department of Atomic Energy (DAE) aimed at development of technologies so that India, over time measured in short decades, can utilise the full energy potential of domestic as well as imported uranium and in the long run, move on to exploit thorium.

The second chapter analyses the global energy landscape and the changing geopolitics of energy markets. The chapter begins with an analysis of the world energy market and the role of major players from both the producers block as well as suppliers block. It argues that the use of various fuel sources vary widely across geographies. However, energy consumption pattern is being increasingly influenced by the evolving technological developments. So far, Europe and North America have been the major beneficiary of the development of nuclear energy sector, the same is now gaining more importance in the Asian region.

The third chapter looks at Asia where nuclear energy sector is expected to experience tremendous growth. The author points to the dichotomy that the advanced societies continue to rely on nuclear power even as emerging economies are facing issues of restrictions on access to nuclear resources and technologies, along with public resistance in their quest for exploitation of peaceful uses of nuclear energy. Development of advanced reactor technologies and their availability will play a critical role in the resurgence of nuclear power in countries like China and India, which together will continue to spearhead the growth of the sector in the region. These countries have also made significant progress in developing nuclear technologies. The indigenously developed technologies will not only better serve the domestic sector by being competitive, but will also have the potential to reach out to other developing nuclear energy markets in future.

One of the most important debates that drives the suitability of nuclear energy is about its potential role in addressing the energy-related emissions. The chapter on nuclear power and climate mitigation examines the relevance of nuclear energy in the clean energy mix of Asian economies. As the developing economies in the globalised world turn out to be the demand centres of energy, it is expected that there can be critical challenges in ensuring supplies with the currently available sources and technologies. While the overreliance on fossil fuels has already signalled alarm among policy makers with regard to the increase in emissions, low-carbon energy mix will be the policy priority. The chapter concludes that nuclear energy will not only be able to meet the increasing demand, but also will be the most suitable energy supply source especially for the high energy demand areas.

In the next chapter, the author argues about the need to rethink the role of the conventional, large, resource-intensive power plants and look at novel solutions. Small and modular reactors (SMR) provide a window of opportunity to realise the potential of nuclear energy, while also addressing issues on safety, waste handling and land footprint in a concerted manner. More importantly, the ever-elusive promise of electrification for all communities spread far and wide (a feature of rural Asia) has remained unfulfilled. This chapter provides an overview of the opportunities for and challenges that are likely to be faced by SMR proponents. The author advocates that Asia must stay connected to the developments to embrace or reject the SMR technology after fully understanding its merits.

The public perception is a critical determinant in the growth of nuclear energy sector. The chapter on risk perceptions specifically focuses on the divergence of perception by experts and by the public. This chapter also explores a full range of alternative perceptions from various stakeholders on the risks associated with the nuclear power sector. Some of the critical nuances about public perceptions relevant to nuclear energy debate are unveiled in this chapter. This will give the readers an in-depth understanding about the underlying factors that shape the differences in perceptions between the experts and the general public.

The chapter on public perceptions of nuclear energy in the post-Fukushima period, examines the response of public and the factors that shape the perception of nuclear risks. The author argues that post-Fukushima, the debate on nuclear safety demands greater attention from scientific community as well policy makers. It further highlights that enhanced risk communication need to be prioritised in societies where nuclear energy exists as a component in the energy mix.

The global nuclear industry has decades of experience in operation and maintenance. The management of this knowledge is a key to preserve as well as transfer information efficiently to the newer generations of nuclear sector professionals. The chapter on nuclear knowledge management focuses on the existing and emerging trends in the global nuclear sector and the potential lessons that can benefit the nuclear energy industry in the Asian region.

The legal aspects related to nuclear issues deserve greater attention in today's world. The nuclear liability and regulatory aspects in the context of South Asian region are explored in detail in the next chapter. The fact that two of the non-NPT states in the world are from South Asia makes the debate on nuclear law interesting. The chapter provides background of nuclear energy programmes in South Asia; tracks the legal, liability and regulatory regimes that follow international law; details individual countries liability and regulatory regime and concludes with a view that SAARC should work with a common approach with respect to nuclear safety and security.

Measuring the costs and benefits of various energy sources in terms of larger politico-security considerations and environmental context, is a complex effort. Coal being one of the cheapest fossil fuel sources is widely being used in many major economies. Other fossil fuel sources also enjoy key positions in the global energy mix. In the last chapter, the authors focus on the critical issue of the cost of nuclear energy and its implications for future penetration of nuclear energy. Conventionally, the cost of nuclear technology is reported as the overnight construction cost (OCC). The authors argue that when looked at from a the holistic perspective, costs should include other external costs (including risks, co-benefits and co-costs) as well as macro-economic costs such as those incurred for climate change mitigation. Analysis reported in this chapter seeks to inform the nuclear debate in Asia from an analytical viewpoint including the implicit risks and costs that underlie the moral positions on this technology.

This book thus presents a comprehensive view regarding various issues related to nuclear energy sector in the Asian region. The arguments presented in various chapters bring in two important sets of points of policy relevance. First, in Asia's future energy mix, nuclear power deserves an indispensable role as a cleaner source of energy. Second, with the strengthening of the institutional base and legal structures, the Asian region will lead the development of nuclear energy industry.

New Delhi, India — Nandakumar Janardhanan
New Delhi, India — Girijesh Pant
Mumbai, India — Ravi B. Grover

Contents

Editors and Contributors

About the Editors

Nandakumar Janardhanan teaches in the Energy Studies Programme of Jawaharlal Nehru University (JNU), New Delhi, India. He specialises in energy and climate policy. The areas of his work include nuclear power, energy security and geopolitics, climate policy, low-carbon development and renewable energy. He has worked with several policy think tanks and research organizations of international repute. Dr. Janardhanan was a Fellow with the 'Climate CoLab' of Massachusetts Institute of Technology (2014–2015) and one of the lecturing staff for the Nuclear Energy training conducted by the International Atomic Energy Agency (2011–2014). He has been associated with the Institute of Chinese Studies (India) as Adjunct Fellow, as Member of Teaching and Advisory team at Strathclyde Centre for Environmental Law and Governance of the University of Strathclyde (Scotland) and as Fellow with the Institute for Global Environmental Strategies (Japan).

Girijesh Pant is former Dean, School of International Studies, Jawaharlal Nehru University (JNU), New Delhi, India. After receiving his Ph.D. from the JNU, he began his career with the Federation of Indian Chambers of Commerce and Industry (FICCI), New Delhi. He then joined the JNU and taught for more than three decades. He was Vice Chancellor of the Doon University, Dehradun, India and GGD University. His areas of research interest include political economy of development, West Asian economies, international energy affairs and India's energy security. He has been a senior Fulbright Visiting Fellow at the University of Illinois and has written extensively and travelled widely. He has been awarded the Lifetime Achievement Award by the Centre for Southeast Asian and Pacific Studies, Sri Venkateswara University, Tirupati, India. He has been associated with the UGC Area studies programme, Indian Council of World Affairs, Indian Academy of Social Science and many non-governmental organizations (NGOs).

Ravi B. Grover occupies Homi Bhabha Chair instituted by the Department of Atomic Energy (DAE) and is a member of the Atomic Energy Commission. He studied mechanical engineering at Delhi College of Engineering, Delhi University, nuclear engineering at Bhabha Atomic Research Centre (BARC) and received a Ph.D. from the Indian Institute of Science, Bangalore.

He has been working in the DAE for about four and a half decades and was Principal Adviser during October 2010 to February 2013. As a nuclear engineer, he has worked in the areas of fuel and core thermal hydraulic analysis, process safety and plant dynamic analysis, process design of primary systems and equipments, and studies related to growth in energy requirements in India.

He conceptualized the setting up of the Homi Bhabha National Institute as a university level institute and concurrent with other responsibilities, he led it for about eleven years until 2016. He participated in negotiations with other countries and international agencies leading to opening of international civil nuclear trade with India.

He was conferred with a Padma Shri in 2014. He is a fellow of the Indian National Academy of Engineering, and the World Academy of Art and Science (Prof. Grover can be contacted at: rbgrover@hotmail.com).

Contributors

Vaibhav Chaturvedi Council on Energy, Environment and Water (CEEW) New Delhi, India

Karthik Ganesan Council on Energy, Environment and Water (CEEW) New Delhi, India

Ravi B. Grover Homi Bhabha National Institute, Mumbai, Maharashtra, India

Nandakumar Janardhanan Energy Studies Programme, Jawaharlal Nehru University, New Delhi, India

Tatiana Karseka International Atomic Energy Agency (IAEA), Vienna, Austria

Girijesh Pant School of International Studies, Jawaharlal Nehru University New Delhi, India

Kapil Patil Indian Pugwash Society, New Delhi, India

M.P. Ram Mohan Indian Institute of Management, Ahmedabad, India

Priyadarshi R. Shukla Indian Institute of Management Ahmedabad, Ahmedabad, India

A. Vinod Kumar Institute for Defence Studies and Analyses, New Delhi, India

Catherine Mei Ling Wong University of Luxembourg, Esch-sur-Alzette, Luxembourg

Yanko Yanev Nuclear Knowledge Management Institute, Vienna, Austria

Part I
Resurgence

Chapter 1
Resurgence of Nuclear Power in India

Ravi B. Grover

1.1 Introduction

Statistics published by the Ministry of Statistics and Programme Implementation (MoSPI 2015) indicate that installed electricity generation capacity in India increased from 145,755 MW as on 31.03.2006 to 284,634 MW as on 31.03.2015, registering a compound annual growth rate (CAGR) of 7.72%. Electricity generation from utilities[1] has seen an increase from 623,819 GWh in 2005–2006 to 1,022,614 GWh in 2013–2014. This amounts to a CAGR of 6.4%. In 2013–2014, additional 156,642 GWh was generated by non-utilities.[2] Growth in both installed capacity and generation has continued apace during 2015–2016 as well (CEA 2016). Growth in electricity generation will continue during coming decades to support economic growth and provide electricity to millions of households in the rural areas, who at present have no electrical connection. For the purpose of comparison, Table 1.1 gives per capita electricity consumption in the world, OECD countries and select countries in the neighbourhood (IEA 2015).

From Table 1.1, one can infer the following: (i) India's per capita electricity consumption is below the world average and well below the average of OECD

[1]Utilities are undertakings whose primary purpose is production, transmission and distribution of electric energy. These could be public sector undertakings, private companies, cooperative organisations, local or regional authorities or governmental organisations.

[2]A non-utility is an independent power producer, which is not a public utility, owning facilities to generate electric power for sale to utilities and end users. Non-utilities may be privately held facilities, corporations, cooperatives, such as rural solar or wind energy producers, and non-energy industrial concerns having capacity to feed excess energy into the grid.

R.B. Grover (✉)
Homi Bhabha National Institute, Mumbai, Maharashtra, India
e-mail: rbgrover@hotmail.com

N. Janardhanan et al. (eds.), *Resurgence of Nuclear Power*,
https://doi.org/10.1007/978-981-10-5029-9_1

Table 1.1 Average per capita electricity consumption in the world, OECD and select countries

Country	Population (million)	Electricity consumption (TWh)	Per capita electricity consumption kWh/capita
World	7118	21,538	3026
OECD	1261	10,179	8072
India	1250	978.82	783
Malaysia	29.72	132.95	4474
Singapore	05.40	47.73	8840
Sri Lanka	20.48	10.82	528
Thailand	67.01	166.66	2487

Source IEA (2015)

countries; (ii) countries such as Malaysia and Thailand, having climatic conditions similar to India, have significantly higher electricity consumption; and (iii) Singapore, another country having tropical climate, has electricity consumption a little higher than the average of OECD countries. There are other countries in the region and their electricity consumption is lower than India, and as an example, data for only one country, namely Sri Lanka, is given. As economy grows, electricity consumption in India will grow and match the level of consumption in the developed world. Extrapolating electricity generation by utilities in 2013–2014 at a CAGR of 6% indicates that the generation in 2049–2050 will be above 8330 TWh. Population of India continues to grow, and for the purpose of demand projections, one can assume it to be about 1.6 billion by the middle of the century. With this assumption, generation of 8330 TWh amounts to per capita generation of about 5000 kWh. A similar conclusion was derived by the author in an earlier paper (Grover and Chandra 2006) as well and is consistent with conclusions included in the report on the "Integrated Energy Policy" issued by the Government of India (Planning Commission 2006).

Looking at ongoing emphasis on improving energy intensity of economy or stated simply energy conservation, a per capita consumption of 5000 kWh per capita per annum could be a reasonable aim for India to achieve by the middle of the century. One can also assume that India will be able to bring down transmission and distribution losses from the present high level of about 22% to the level of global average of about 7%. Considering this data (5000 kWh per capita consumption, 1.6 billion people, 7% losses), demand for electricity in India will be about 8600 TWh of electricity per annum.

The narrative above is an attempt to arrive at an estimate for the future growth in electricity generation based on an extrapolation of historical data. It can also be looked at from the perspective of human needs, and this is done in the next section.

1.2 Need for Electricity

The question "how much energy or electricity is enough to satisfy human needs" is a difficult question. Its difficulty arises from the fact that there cannot be a universal definition of "human needs". Needs are continuously evolving as the society is moving further and further from the basic hunter-gatherer existence and aspirations are rising.

1.2.1 Early Estimates

In 1971, Scientific American published an article titled "The Flow of Energy in an Industrial Society" by a geo-scientist Earl Cook, wherein he writes how much energy humans consume in the form of food, to build habitat, to move around, to control indoor climate, etc. (Cook 1971). This article is a much referenced article in thinking about how energy usage has evolved ever since humans started inhabiting earth and has been used by Ian Morris to develop theories about relationship between energy usage and social development (Morris 2010) (Table 1.2).

During the hunter-gatherer period, energy was needed only for supporting biological life of individuals that is for endo-somatic metabolism. For a human being to survive about 2000 kcal per day is enough (this translates to 850 kWh per annum[3]). Some energy, of course, was needed for cooking and building a shelter. The need for food could be met in an inexpensive manner by food grains and vegetables or by in an expensive manner by animal meat. Energy consumption increased with the spread of agriculture and use of domestic animals in farms. Increase in energy consumption continued with the spread of advanced agricultural practices. Advent of technology revolution made energy available at large-scale, resulting in further technological development. This included developments in health sciences, increasing longevity and hence increasing population. As population and population density (number of people per square kilometre) increased, agricultural productivity increased to provide food for all. Increase in agricultural productivity was made possible by the use of energy-intensive farming techniques. In highly technological society such as the USA, per capita energy consumption as estimated by Earl Cook in 1971 was 230,000 kcal, which is equivalent to a consumption of 8.03 toe.[4] As per Key Energy Statistics, this figure in 2013 was 6.92 toe for the USA and 4.20 for OECD countries and the world average was 1.90 (IEA 2015). For India, the corresponding figure is 0.62 toe.

[3]1 kWh = 860.421 kcal; 1 kcal = 4.184 kJ.

[4]1 toe = 10 million kcal.

Table 1.2 Daily per capita energy consumption in 1000 kcal during various stages of human development

	Food: plant and animal including animal feed	Home and commerce	Industry and agriculture	Transportation	Total
Technological man	10	66	91	63	230
Industrial man	7	32	24	14	77
Advanced agricultural man	6	12	7	1	26
Primitive agricultural man	4	4	4	–	12
Hunting man	3	2	–	–	5
Primitive man	2	–	–	–	2

Source As estimated by Cook (1971)

1.2.2 *Human Development Indices Versus Per Capita Electricity Consumption*

Many researchers have tried to calculate the energy as well as electricity needs by plotting energy or electricity consumption versus per capita GDP. Several alternate plots such as Human Development Index or life expectancy at birth or some other indices (Lambert et al. 2014; Gopinath 2007) versus electricity or energy consumption have also been used with the objective to draw conclusions about how much energy or electricity is needed for a country. All such plots lead to conclusion that after some level of electricity consumption, any further increase in electricity consumption results in diminishing returns in terms of any of the indices. Based on extensive analysis, Alan D Pasternak concludes that electricity consumption beyond 4000 kWh per capita per annum provides only marginal improvement in HDI (Pasternak 2000). Lambert et al. conclude that countries with per capita energy consumption of 100–200 GJ per annum have a high standard of living and no significant improvement in well-being is seen when per capita energy consumption exceeds 200 GJ per annum. Gopinath writes that beyond per capita consumption of 15,000 kWh per annum, there is no impact on any of the human development indicators. Essentially, all studies point to a saturation effect with no improvement beyond a certain level of energy or electricity consumption.

The approach based on correlating various indices to quantify need for electricity is not shared by many others. Andrew Teller reviews the paper by Pasternak and similar other papers and says that "correlation is not causation" (Teller 2009). He writes, "Apart from the fact that the widespread use of electricity undoubtedly contributes to a population's wellbeing, it is also a sign of capability of this population of engaging in fairly high-tech activity. Generating and distributing electricity requires skilled personnel, hence a high literacy rate; it requires large investments, hence a sufficiently high GDP, which in turn is needed to enable the consumers to pay their bills; finally, it is best ensured in periods of peace, which in turn promotes schooling and ensures higher life expectancy. Seen in this light, HDI and electricity consumption appear to be two facets of a more general phenomenon, i.e. the capability of a society for generating sufficient income and using it to the benefit of its members".

The author tends to agree with the opinion that correlation cannot be equated to causation. The exact value of annual per capita electricity consumption needed by a country depends on several socioeconomic factors, which will keep evolving as technologies (both for energy generation and for energy use) are developed and societies evolve. Evolution includes aspiration of human beings to spend resources on higher needs such as art, culture and other leisure time activities. Moreover, electricity is the best energy carrier from the point of end-use convenience, and its share in total energy consumption will rise with improvement in technology. According to an IAEA report (IAEA 2015), worldwide primary energy used for electricity in 2014 was 37.3% and it is estimated to go to 47% in 2050. Analyses based on HDI-kWh/capita plane cannot capture such changes.

1.2.3 Estimates for India

While attempts have been made to benchmark annual per capita needs of electricity in India (Sukhatme 2011), in the opinion of the author, one should not attempt to define need of energy only in terms of electricity. Per capita energy consumption in any country including India will be a dynamic number and change with parameters such as share of electricity in the energy mix, technological advances, density of population, level of education, income that can be spared for higher needs (art, culture and leisure time activities) and many others. Additionally, in a globalised world, one cannot expect people in India to be satisfied with 2000 or 4000 kWh per annum per capita, when average of OECD countries is 8000 kWh per annum per capita and countries in the immediate neighbourhood have higher electricity consumption. In view of this, it is best to use the earlier narrative based on extrapolation of historical trends to come up with an estimate of per capita electricity growth and arrive at an aim which India should achieve by the middle of the century. As stated in the previous section, based on extrapolation of historical data, aiming to achieve per capita electricity consumption of 5000 kWh by the middle of the century is a feasible aim and can be a basis for planning the growth of electricity generation in India.

1.3 Supply-Side Perspective

Narrative in the previous sections brings out how the demand for electricity will increase in the coming decades, and this section examines the supply options available to meet the demand. The demand has to be met in a manner that takes into consideration available energy resources, environmental sustainability, affordability and technological feasibility.

1.3.1 India's Energy Resources

India is not rich in energy resources. Three main contributors to electricity generation are thermal (predominantly coal fired) power stations, nuclear power stations and large hydro power plants. Contribution from renewable sources such as solar, wind and small hydro is growing. Coal is likely to be the mainstay of electricity generation in the coming decades. As on 31.04.2014, the estimated total coal reserves were 301.05 billion tons, 42% being proven, 47% indicated and the rest inferred (MoSPI 2015). All proven coal[5] reserves are not likely to be economically mineable. Production of coal in the year 2013–2014 was 565.77 million tons, and additionally, 166.29 million tons were imported. All efforts are being made to increase domestic production so as to bring down coal imports. Considering growing needs of coal by electricity generation plants and other users, economically mineable coal reserves are not likely to last beyond a part of this century. Hydrocarbon reserves are very modest (762.74 million tons of crude oil and 1427.15 billion cubic metres of natural gas as on 31.03.2014), and these are predominantly used for other sectors of economy such as transport and fertilisers, etc. India has to import large quantities of hydrocarbons to meet all requirements, and hydrocarbons cannot be a significant resource for generation of electricity. Among fossil fuels, coal has been and will continue to be the mainstay for power generation in India.

A ranking study of hydroelectric schemes in India done by the Central Electricity Authority (CEA 2002) has assessed the total hydroelectric potential at 84,044 MW at 60% load factor which will produce 442 TWh of electricity. This potential when fully developed would result in an installed capacity of 150,000 MW on the basis of a probable average load factor. As on 31.03.2016, installed hydro capacity was 42,783 MW, and in the year 2015–2016, generation by hydro plants was 121 TWh (CEA 2016). A lot needs to be done to exploit full hydropotential in the country, but considering issues involved, particularly displacement of people, it is taking longer than envisaged. A group set up by Niti

[5]Lignite is also used in India for the generation of electricity. To keep the argument simple, lignite resources, etc., are not included in the narrative here.

Aayog is working out India Energy Security Scenario 2047 (IESS 2015).[6] A recent document by IESS gives various scenarios for full exploitation of hydropotential, and the most optimistic scenario (christened "heroic effort scenario") envisages complete exploitation only by 2047 (IESS 2015). The scenarios are still being evolved, and one should take this statement as a work in progress.

India has made significant progress in harnessing nuclear technology for power generation, but growth in installed nuclear capacity has been slow to pick up due to inadequate production of uranium in the country. This changed in 2008 with the opening up of international civil nuclear trade, and availability of uranium is no longer an impediment in harnessing nuclear technology for power generation (Grover 2015). The Department of Atomic Energy has ambitious plans for short-term as well as long-term growth of installed nuclear capacity. This is detailed later in this chapter.

The total potential for renewable power generation, other than solar, in the country in terms of installed capacity has been estimated at 147,615 MW as on 31.03.14; this includes wind (power potential at 102,772 MW), small hydro (power potential at 19,749 MW), biomass (power potential at 17,538 MW) and bagasse-based cogeneration in sugar mills (power potential 5000 MW) (MoSPI 2015). Precise data about potential for actual generation is not available, but renewable sources work at low capacity factors. Assuming an average capacity factor of 20%, total electricity that can be generated from renewable sources other than solar is about 250 TWh. Even if one adds about 500 TWh as the likely contribution from solar, the total number is not a significant percentage of what India has to aim to generate by the middle of this century. However, considering the distributed nature of renewable sources and their potential for providing electricity to remote areas, there is a strong case for harnessing the full potential of all renewable sources. Technologies for harnessing renewables are continuously evolving, and it is difficult to precisely predict the future in this area. Ongoing research and development, particularly in the area of concentrated solar power, can be a game changer.

1.3.2 Environmental Sustainability

Projected level of electricity generation by the middle of this century represents a sevenfold increase over what was generated in India during the fiscal year ending March 2016. This is a massive increase and continued reliance on fossil-fired fuels to meet increasing demand will have impacts in several areas including

[6]The IESS, 2047, has been developed expressly as an energy scenario building tool. The guiding ambition of IESS is to develop energy pathways leading up to the year 2047, comprising likely energy demand and supply scenarios. The data relating to implications—energy security, costs, land and carbon dioxide emissions—are merely indicative and not firm estimates. Therefore, this is a scenario building exercise and not strictly an energy model.

environment and human health. Globally, fossil fuels have come under attack for their environmental impact, and there is a call for their phase-out. Arguments are being advanced for and against nuclear and also some forms of renewable sources. Mix of technologies to be used for electricity generation is being debated in several forums based on "impressions and emotions" as well as based on "an evidence-based approach". In a landmark study, Brook and Bradshaw (2015) argue that "the best energy options are those that use the least amount of land and fresh water, minimise pollution, restrict habitat fragmentation and have a low risk of accidents that have large and lasting impacts on natural areas". Energy sources, of course, have to be cost-effective. They have ranked seven different options for electricity generation based on seven select indicators. Indicators selected by them are greenhouse gas (GHG) emissions, electricity cost, despatchability, land use, safety, solid waste generation and radioactive waste generation. GHG emissions were tabulated in tons of carbon dioxide per TWh and include both emissions during production and plant life cycle. Cost of electricity includes cost amortisation for long-term waste management and decommissioning for nuclear power plants. Despatchability represents capacity and availability to deliver electricity on demand. Safety is represented by deaths from accidents and excludes chronic health effects. Land use includes land required for generation and fuel mining. Energy resource with the lowest environment or economic impact indicator is assigned a rank of one, and the one with the worst indicator is assigned the rank seven.

Considering ranks 6 and 7 as poor, solar PV has poor rank for cost and despatchability. Further, at the present level of deployment of solar PV, data regarding deterioration of performance over time in tropical climate having heavy dust load is scarce and there is no clarity regarding problems that could arise from disposal of used PV cells on a large scale. Another important issue associated with solar is its poor value of the ratio of energy returned on invested (EROI), and this is discussed in detail in a later section. Wind has poor rank for despatchability due to its intermittent nature, biomass and coal because of several parameters, and hydro for land use. Natural gas has poor rank only for GHG emissions.

All ranks were combined using a multi-criteria decision-making analysis framework. The final weighted rankings of various methods of electricity generation were as follows: nuclear (1.3), natural gas (2.0), onshore wind (2.3), hydro (3.3), solar PV (5.3), coal (6.0) and biomass (6.7). The study by Brook and Bradshaw does not rank concentrated solar power (CSP) and offshore wind. The study also does not consider chronic health problems and EROI. Health effects arise due to pollutants such as particulate matter and harmful gases released through the stack and manifest as external costs.

The term external costs is used to denote the cost that the party responsible for generating emissions does not account for and, consequently, consumers of electricity do not pay for. External costs are paid in terms of health effects (deaths, serious illness, minor illness) by those who are exposed to emissions. External costs have been extensively studied in the European Union through ExternE project over about 15 years' period. Study follows dose–response approach, where pathways through which pollutants are dispersed are mapped, dose of pollutants received by humans is estimated, its health effects are studied, and finally, a monetary estimate (lost working days and cost of medication) of the health effect is evaluated. Results are well summarised by Markandeya and Wilkinson (2007), and nuclear has minimum health effects or external costs among the electricity generation technologies studied, namely lignite, coal, oil, gas, biomass and nuclear. Renewable sources have environmental advantages, but have their drawbacks too. Markandeya and Wilkinson bring out four main drawbacks: low energy density, intermittency, constraints on their location and environment issue (including aesthetic effects), which might in part offset part of their environmental and health gains derived from lower air pollution and greenhouse gas emissions.

Overall from the point of environment sustainability, one has to increase contribution of nuclear and renewables in the electricity mix. One also has to examine various technology options from consideration of economics or affordability.

1.3.3 Affordability

A comparative study of economics of electricity generation by various competing technologies has to consider several aspects: the capital cost of setting up of a power plant; the generation cost which combines the cost of capital (that is return on equity and interest on debt), fuel, operation and maintenance; external costs referred to earlier; and finally, net energy gain represented by the ratio of energy returned to energy invested (EROI). Capital costs and generation costs are important from commercial consideration and so historically have been the focus of businesses as well all studies. External costs and net energy gains are important for society, but have not received the same level of attention. External costs came into sharper focus towards the end of the previous century, and as explained in the previous section, ExternE project funded by the European Union led to the development of a rigorous methodology. The concept of net energy gain is a defining characteristic for comparison of energy supply options for society as a whole and has been around for a few decades. In the 1960s and 1970s, it was articulated in terms of life cycle assessments, in the 1980s in terms of net energy gain and more recently in terms of EROI (Raugei and Leccisi 2016; Weisbach et al. 2013; Hall et al. 2014). Essentially, this is a concept that brings out the characteristic that irrespective of the energy carrier used, at the end there should be a net energy gain for the society. For calculating EROI, one has to make lifecycle assessment for the energy-generating technology.

A number of definitions for EROI and methodologies for its evaluation have been proposed by different researchers and are in use. As a result, the values of EROI for various electricity-generating technologies calculated by researchers differ markedly. One important factor for the difference is the approach adopted towards adding primary energy and electrical energy. Energy input to an electricity-generating technology can be a mix of both primary energy and electrical energy. One can simply add them or one can convert electrical energy to “energy content” of electricity following the convention[7] used by the International Energy Agency (IEA 2015) or similar other convention before adding. Second reason pertains to defining the boundary of the system being analysed. Take the case of a light water reactor using enriched uranium as fuel. If the enrichment plant is considered to be co-located with the nuclear power plant and electricity used by it is subtracted from the output of the nuclear power plant, one gets one result. If the enrichment plant is considered to be outside the system boundary and consequently the electricity used by the enrichment plant is taken as a component of the energy invested, one gets a different result.

For calculating EROI for an electricity-generating plant for a country, when a part of input is imported, one has to use energy invested in earning the capital used for imports.

In spite of difference in approach for evaluating EROI, one can analyse various studies to get insight about net energy gain of different technology options and draw conclusions with regard to their usefulness for the economy of a country. Such an analysis brings out that (i) coal, nuclear and hydro have high values of EROI; (ii) biomass has a low value of EROI; (iii) concentrated solar power has higher EROI as compared to solar PV and, within solar PV, there is a difference between technologies. A vast majority of PV systems use crystalline silicon modules, while ground-mounted cadmium telluride (CdTe) thin-film PV installations are also appearing. CdTe-based PV has superior EROI as compared to silicon-based PV installations. The case of oil is very complex as the EROI for the production of oil is continuously declining. It has decreased from a high value of above 1000 in the early decades of the twentieth century to low double digit numbers now, and sources such as tar sands and oil shale have an EROI of less than 10 (Guilford et al. 2011; Hall et al. 2014). However, technology for exploiting oil shale is continuously evolving, and one can see improvements in coming years. It is difficult to get EROI for gas production alone as it is associated with oil, but gas-fired power plants have high thermodynamic efficiency which is a contributor towards increasing their EROI.

As stated earlier, during the hunter-gatherer period, energy was needed only to support the needs of the human body, that is, endo-somatic metabolism. As the society developed, needs for energy increased continuously to support complex metabolism of the modern society which one may call as exo-somatic metabolism (Raugei and Leccisi 2016). Energy used has become a measure of the real amount of work done by the society, and it is necessary that energy sector of a society has sufficiently high EROI so that it supports not only itself, but also all other non-energy-producing sectors of the economy. This makes it necessary to ensure that high EROI sources form a sufficient part of the energy mix of a nation. Excessive reliance on renewable sources can bring down the EROI and so is not

[7] Hydro-electricity production and electricity produced by other non-thermal means (wind, tide/wave/ocean, photovoltaic, etc.) are accounted for by using 1 TWh = 0.086 Mtoe. However, the primary energy equivalent of nuclear electricity is calculated from the gross generation by assuming a 33% conversion efficiency, i.e. 1 TWh = 0.086–0.33 Mtoe.

desirable. Coal has high EROI, but has deleterious consequences on environment when its use increases in a developed society. In an underdeveloped economy, where the society relies on traditional biofuels, the use of coal replaces biofuels and that is a net gain for environment and such use will lead to development of the society.

Therefore, as India develops, it has to ensure that its electricity mix is so engineered that it has high enough EROI. What EROI is enough for the grid of a nation and what EROI is necessary for a source to be viable? While some estimates have been proposed, in the opinion of this author, a precise answer is difficult as methodology for estimating EROI is yet to be standardised. One can, however, draw qualitative conclusions based on available estimates. To quote data from Weisbach et al., electricity-generating technologies having an EROI above 10 are as follows: nuclear[8]—105; hydro—49; coal—30; combined cycle gas turbine—28; concentrated solar power—19; and wind—16. Weisbach et al. report low values for solar PV, but other publications indicate that EROI for solar PV is improving and is above 10 for certain types of PV cells (Bhandari et al. 2015). Bhandari et al. use a methodology which is different[9] from that used by Weisbach et al. and calculate an EROI which is high enough to make them viable. All estimates point towards a low value of EROI for biofuels.

From the spread of the value of EROI of various technologies, one can infer that for the grid to have a high EROI, continued use of coal is necessary until contribution from clean sources such as nuclear and renewables can be increased in a manner that overall EROI remains high. Nuclear is important not only from the point of environmental sustainability, but also from an economic standpoint as it has high EROI and low external costs. Therefore, from a long-term point of view, considering economics and environment benefits, one has to aim at a mix of renewables and nuclear.

1.3.4 Technological Issues

Continuous advances in technology are helping to resolve technological issues, but ever-increasing scale of deployment of electricity-generating technology is throwing up new challenges. For example, the use of coal as a replacement for unscientific burning of biofuels is clean, but its large-scale use has given rise to an increase in concentration of greenhouse gases in the atmosphere and resulting global warming. Scientific community has to find a solution for this problem, and advanced coal technologies have been successfully developed to reduce emission of

[8]It assumes that uranium is enriched in a plant operating on centrifuge technology. In case diffusion technology is used, EROI comes down to 75.

[9]Bhandari et al. convert all energy invested and output to primary energy equivalent by using primary to electrical energy conversion efficiency using average number for the country's grid. Weisbach et al. add electrical energy and primary energy directly.

carbon dioxide. Carbon capture and storage (CCS) techniques have been developed, but are yet to be tried on a large enough scale to provide confidence in their use. Research is also needed to ensure that large-scale storage does not give rise to any safety issues. At the current stage of development of CCS technologies, there could be unknowns and more research is needed to identify and resolve all possible issues. Resolution has to be so done that it does not have a significant influence on EROI.

Portfolio of technologies for exploiting renewable sources is very wide. Common issues are their intermittent nature, constraints on their location and low energy density. Intermittent nature poses challenges for integrating renewables in the grid and imposes large system cost for integration. System cost has several components: provision of operative reserve[10] (spinning reserve, supplemental reserve), replacement power[11] and upgradation of power grid to accept dynamic power input from many decentralised and distant variable sources. This could change if ongoing research is successful in developing low-cost storage technologies. Large-scale usage of renewables can bring up some unknowns, and they will have to be addressed. Low EROI is an issue for biofuels and solar PV. Research is needed to come up with better materials for manufacturing solar PV films that are more efficient in converting solar energy into electrical energy. Concentrated solar power holds promise and needs to be tried on a large scale.

Fission-based nuclear reactors have been around since the middle of the previous century, have been deployed on large enough scale for providing base-load electricity and have been subjected to much deeper scrutiny with regard to safety than any other modern technology. Nuclear fuel has high energy density, and nuclear power plants have large enough EROI, lowest external costs, low GHG emissions and competitive generation costs. As electricity usage increases, it is desirable to attach monetary value to all parameters influencing society so that all energy-generating technologies are treated on a level playing field.

Choice of nuclear fuel cycle has significant economic as well as sustainability implications. The fuel cycle can be an open fuel cycle that is fuel once used is stored for disposal; it can be a partially closed fuel cycle where fuel discharged is reprocessed once and recovered plutonium is used as mixed oxide fuel (MOX) in LWRs; it can be a fully closed fuel cycle where used fuel is subject to multiple

[10]Operative reserve in a grid is the generator capacity available to the grid manager to meet demand and supply fluctuations. Intermittency of renewable sources results in large variations on supply side. Spinning reserve is the extra generation capacity that is available by increasing the output of generators that are already connected. Such a supply can come on line within seconds to minutes. Supplemental reserve is the extra generating capacity that is not currently connected and can be brought on line after a short delay. It could be from fast start generators or from an interconnected grid.

[11]Replacement power is reserve power provided by generators requiring long start-up time, several minutes and even going to an hour. It relieves generators providing spinning and supplemental power, thereby restoring operating reserve in the grid. Intermittence of renewable implies that replacement power that is equal to installed renewable capacity has to be available and that has to come from base-load stations.

recycling in fast breeders so that full energy potential of heavy metal is realised; or it can be fully closed with actinide partitioning and transmutation. The fourth option that is a fully closed fuel cycle with multiple recycling in fast breeders, actinide partitioning of high-level waste and transmutation of separated actinides is the way to go in the future. By following this approach, one is able to address the issue of radioactive waste disposal and all technology elements are in place to take this route. While India has decided to follow this approach[12] (Grover 2015), it needs to be followed at the global level. In many other countries, while scientists are convinced about it (Hannum et al. 1997), political will is lacking. Pursuit of closed fuel cycle will also enable use of thorium for power production, which is more abundant than uranium. It may be added that closed fuel cycle improves renewability of nuclear power and increases its EROI. Uranium is available in sea water, and ongoing research to extract this uranium can provide further fillip to availability of uranium and nuclear power.

1.3.5 Role of Nuclear in the Electricity Mix

One can conclude this section by saying that considering all aspects, nuclear power has to be a significant part of India's energy mix. Renewables should be exploited to the extent possible, and coal cannot be wished away in the near future. In the Indian context, coal replaces traditional biofuels, which are more damaging to environment. Increased reliance on nuclear power is necessary at the global level as well to stabilise global temperatures before the year 2100. Achieving a large enough deployment of nuclear power by the year 2100 is possible if one were to rely on deploying breeder reactors, increase contribution from Pressurised Heavy Water Reactors (PHWRs) and deploy thorium-fuelled reactors. Led by Harve Nifenecker, several individuals including this author came together and have built a scenario to achieve such an objective (Berger et al. 2016). To critics of fuel reprocessing and breeder reactors, one would like to say that it is much safer to use plutonium as fuel in breeder reactors than to store it as a part of the spent fuel (Grover 2014a). World Nuclear Association, a nuclear advocacy group based in London and having global membership, has

[12]Closed fuel cycle produces minimum waste per unit of electricity generated and also ensures sustainability of uranium by using its full energy potential. Spent fuel consists of fission products and minor actinides, and uranium and plutonium. Reprocessing recovers uranium and plutonium. Minor actinides have a very long half-life, and their presence in the waste makes it necessary to store the waste for a long time. Nuclear waste can be partitioned, a process that separates minor actinides from the waste. The remaining waste then would require storage for a period of about 300 years, which is a historical time frame. Minor actinides can be fabricated into fuel and burnt in fast reactors. The process of burning is called transmutation. An engineering-scale demonstration facility for partitioning of minor actinides has been set up in India and is already working. Prototype Fast Breeder Reactor (PFBR) is nearing completion, and the step remaining to solve the waste problem is to demonstrate the process of transmutation. Partitioning of waste from fast reactors will also require similar demonstration.

proposed what they have christened vision *Harmony* targeting a contribution of 25% of electricity by nuclear by the year 2050. This will require construction of 1000 GW of new nuclear capacity and have proposed a pathway to build such a large capacity (WNA 2016).

1.3.6 Public Perception

In spite of nuclear experts explaining advantages of nuclear energy based on facts and figures pertaining to environment sustainability, economics including external costs as well as EROI, a safety record that is better than all other energy-generating options, and scientific solutions for waste disposal, lay person is not convinced about the safety of nuclear power and there is a sharp divide between “experts” and “lay persons”. Human sense organs are equipped to sense sight, smell, taste, touch and sound and are not equipped to sense the presence of radiation. Therefore, human beings have to depend on instruments to detect the presence of radiation and its level. The dependence on instruments, which are controlled by experts and not lay persons, perhaps leads to distrust about radiation. Anyone speaking in favour of nuclear energy has to answer questions related to accidents at TMI, Chernobyl and Fukushima though when examined in terms of various identifiable parameters including fatalities per TWh of electricity generated as has been done by Brook and Bradshaw (2015) and similar other studies, nuclear energy gets highest ranking. It is time to “weigh up the advantage and disadvantages without recourse to ill-founded ideological radiophobia” (Grimes 2016). Even in situations when the silent majority is in favour of nuclear power projects, vocal minority attempts to influence the decision-making process as happened in the case of the nuclear project at Kudankulam in southern part of India. Nuclear industry has to increase engagement with public and proactively explain all about nuclear energy in a language that public can understand. Winning trust of silent majority and having a reasoned discussion with people opposing nuclear projects based on facts has to be the way forward. Governance systems, including systems for management of disasters, have to be designed to win the trust of common man and appropriate narrative need to be used to convey convincing messages to the public.

1.4 Resurgence of Nuclear Power in India

India has been engaged in development of technologies for exploiting nuclear science and technology since the 1950s and has been successful in mastering technologies as well as setting up infrastructure spanning all aspects of fuel cycle. Extensive facilities for human resource development have been created. In spite of all these positives, share of nuclear generation in the total generation remains low. This is due to poor availability of uranium in the country. During the first decade of this century, domestic uranium supply was not adequate to fuel the PHWRs and capacity factors of operating nuclear power reactors came down to as low as 50% in the year 2008. This changed with the relaxation of guidelines by the Nuclear Suppliers Group in the year 2008 (Grover 2014b, 2015). India is now able to import uranium, and capacity factors of existing reactors have dramatically improved. Now, it is possible for India to set up more PHWRs based on indigenous capability and increase installed nuclear capacity base.

In December 2015, the Government of India submitted to CoP21 its "Intended Nationally Determined Contributions (INDC) for the Period 2021–2030". Through this statement, India has committed "to achieve about 40% cumulative electric installed capacity from non-fossil fuel-based energy resources by 2030 with the help of transfer of technology and low-cost international finance from Green Climate Fund (GCF)". The phrase "non-fossil fuel-based energy" includes nuclear energy.

Based on research done in the research centres of the Department of Atomic Energy, several technology options are now available for an accelerated growth of installed nuclear capacity. Overall, steps taken by the Government have established conditions for the resurgence of nuclear power in India. These steps are the following (Grover In press).

1.4.1 Nuclear Cooperation Agreements

Nuclear cooperation agreements have been signed with several countries including Argentina, Australia, Canada, France, Kazakhstan, Namibia, Korea, Russia, Sri Lanka, the UK and the USA. Agreements with Bangladesh, EU and Japan have been finalised. This will enable India to import uranium from several sources and also set up reactors in technical collaboration with them. Many cooperation agreements provide for joint research and development, which will be mutually beneficial. Consistent with the policy to pursue a closed fuel cycle, all nuclear cooperation agreements include reprocessing consent.

1.4.2 Evolution of Governance System

Governance system in India to support an accelerated growth of nuclear power is being evolved. Recently implemented elements include enactment of a liability legislation to provide prompt compensation to the victims of a nuclear accident, signing and ratifying convention on supplementary compensation, amendment to Atomic Energy Act to permit joint ventures between two public sector companies to set up nuclear power plants, signing and ratifying convention on physical protection on nuclear materials, participating in the Nuclear Security Summit process and working towards strengthening nuclear security architecture. Atomic Energy

Regulatory Board has over the years evolved as a competent regulatory agency and enjoys full functional independence.

1.4.3 Near-Term Growth

Nuclear Power Corporation of India Limited (NPCIL) has extensive experience in design and construction (Bohra and Sharma 2006) of PHWRs and is already constructing four more PHWRs. NPCIL has capability to set up PHWRs at a competitive cost and has plans to set up more PHWRs at existing sites as well as new sites which have also been identified. Sites to set up reactors in technical collaboration with other countries have been identified, and spade work to proceed further has been done. Agreements with Russia to set up more reactors at Kudankulam have been signed, and a second site to set up nuclear power plants in collaboration with Russia is to be announced soon. Good progress has been made in talks with Westinghouse, and key concerns with regard to liability legislation have been addressed (Biswal 2016).

1.4.4 Budgetary Support for Nuclear Energy

Government has made commitment to provide budgetary support to nuclear power as can be seen from the paragraph 82 of the budget speech of the Finance Minister to Parliament on 29 February 2016 which reads, "In the power sector, we need to diversify the sources of power generation for long term stability. Government is drawing up a comprehensive plan, spanning next 15–20 years, to augment the investment in nuclear power generation. Budgetary allocation up to Rs. 3000 crores per annum, together with public sector investments, will be leveraged to facilitate the required investment for this purpose".

1.4.5 Technological Development by the DAE to Ensure Long-Term Growth

A Prototype Fast Breeder Reactor (PFBR) being set up by the Department of Atomic Energy (DAE) is expected to achieve first criticality in 2017, and more such reactors are planned. Exploratory studies to use slightly enriched uranium in PHWRs are being pursued. The DAE has already started work on designing a pressurised water reactor (PWR), and India is participating in the ITER project as a full partner.

1.5 Conclusions

Considering historical trends during the past decade, global average per capita consumption and per capita consumption in India's neighbourhood, it is evident that India will have to plan for an electricity-generating capacity that can provide about 5000 kWh per capita per annum by the middle of the century. Assuming population in the year 2050 to be about 1.6 billion, transmission and distribution losses at 7%, this will call for a generation of about 8600 TWh per annum. This is about a sevenfold increase over the generation in 2015–2016 and calls for careful planning and consideration of all aspects including availability of resources, technologies to exploit resources, environment sustainability and economics including factors such as EROI and external costs which have been hitherto neglected. Nuclear energy, because of its inherent advantages including environmental sustainability and competitive economics when full spectrum of costs is considered, can be a significant part of the electricity mix, and growth in installed nuclear capacity needs to be accelerated. Several initiatives have been taken by the Government towards resurgence of nuclear power, and one can expect to see results within about a decade. Nuclear industry must continuously engage public and address their concerns in a language that is easy to understand.

References

Berger, A., Blees, T., Breon, F. M., Brook, B. W., Hansen, P., Grover, R. B., et al. (2016). How much can nuclear energy do about global warming. *International Journal of Global Energy Issues*, *40*(1/2), 43–78.

Biswal, N. (2016). U.S. official says India has addressed nuclear concerns, as quoted by Reuters. Accessed July 6, 2016 at http://www.reuters.com/article/us-usa-india-nuclearpower-idUSKCN0YF2C1.

Bhandari, K. P., Collier, J. M., Ellingson, R. J., & Apul, D. S. (2015). Energy payback time (EPBT) and energy return on energy invested (EROI) of solar photovoltaic systems: a systematic review and meta-analysis. *Renewable and Sustainable Energy Reviews, 47,* 133–141.

Bohra, S. A., & Sharma, P. D. (2006). Construction management of Indian pressurised water reactors. *Nuclear Engineering and Design, 236,* 836–851.

Brook, B. W., & Bradshaw, C. J. A. (2015). Key role for nuclear energy in global biodiversity conservation. *Conservation Biology, 29,* 702–712.

CEA. (2002). Preliminary ranking study of hydroelectric schemes. Accessed at http://cea.nic.in/reports/others/hydro/hpi/ranking_study/gen_report.pdf July 1, 2016.

CEA. (2016). Executive summary, power sector. Ministry of Power, Central Electricity Authority, March 2016. Accessed June 5, 2016 at http://www.cea.nic.in/reports/monthly/executivesummary/2016/exe_summary-03.pdf.

Cook, E. (1971). The flow of energy in an industrial society. Scientific American, 134–144.

Gopinath, D. V. (2007). Radiation effects, nuclear energy and comparative risks. *Current Science, 93,* 1230–1248.

Grimes, D.R. (2016). Why it's time to dispel the myths about nuclear power. *The Guardian*, April 11. Accesses on 14 July 2016 at https://www.theguardian.com/science/blog/2016/apr/11/time-dispel-myths-about-nuclear-power-chernobyl-fukushima.
Grover, R. B., & Chandra, S. (2006). Scenario for growth of electricity in India. *Energy Policy, 34* (2006), 2834–2847.
Grover, R. B. (2014a). Technological dimension of nuclear security. *Strategic Analysis, 38*(2), 151–156.
Grover, R.B. (2014b). Nuclear power for India's energy security: external and internal challenges. 13th Volume of India's National Security Annual Review published by Routledge, India, 2014. ISBN 978–1-13-879638-6.
Grover, R. B. (2015). The civil nuclear programme: the last decade and the outlook for near-future. *Indian Foreign Affairs Journal, 10*(3), 255–267.
Grover, R. B. (In Press). Opening up of international civil nuclear cooperation with India and related developments. Progress in Nuclear Energy. http://dx.doi.org/10.1016/j.pnucene.2016.09.016.
Guilford, M. C., Hall, C. A. S., O'Connor, P., Cleveland, C. J. (2011). A new long term assessment of energy return on investment (EROI) for U.S. oil and gas discovery and production. *Sustainability, 3*, 1866–1887.
Hall, C. A. S., Lambert, J. G., & Balogh, S. B. (2014). EROI of different fuels and the implications of society. *Energy Policy, 64,* 141–152.
Hannum, W. H., Wade, D. C., Mcfarlane, H. F., & Hill, R. N. (1997). *Progress in Nuclear Energy, 31*(1/2), 203–217.
IAEA. (2015). Energy, electricity and nuclear power estimates up to 2050.
IEA. (2015). *Key world energy statistics.*
IESS. (2015). User guide for India's 2047 energy calculator: large hydro sector. Accessed at http://indiaenergy.gov.in/docs/Hydro%20Power_Documentation.pdf July 1, 2016.
Lambert, J. G., Hall, C. A. S., Balogh, S., Gyoza, A., & Arnold, M. (2014). Energy, EROI, and quality of life. *Energy Policy, 64,* 153–167.
Markandeya, A., & Wilkinson, P. (2007). Electricity generation and health. *Lancet, 370*, 979–990.
Morris, I. (2010). *Social development.* Stanford University, Accessed June 5, 2016 at http://ianmorris.org/docs/social-development.pdf.
MoSPI. (2015). *Energy statistics 2015*, 22nd issue. Central Statistics Office, Ministry of Statistics and Programme Implementation, Government of India, New Delhi.
Planning Commission. (2006). *Integrated energy policy*. Report of the Expert Committee.
Pasternak, A. D. (2000). Global energy futures and human development: A framework for analysis, October 2000, UCRL-ID-140773.
Raugei, M., & Leccisi, E. (2016). A comprehensive assessment of the energy performance of the full range of electricity technologies deployed in the United Kingdom. *Energy Policy, 90,* 46–59.
Sukhatme, S. P. (2011). Meeting India's future needs of electricity through renewable sources (Vol. 101, pp. 624–630).
Teller, A. (2009). *Correlation is not causation*. European Nuclear Society, e-news Summer Issue.
Weisbach, D., Ruprecht, G., Huke, A., Czerski, K., Gottlieb, S., & Hussein, A. (2013). Energy intensities, EROI (energy returned on invested), and energy payback times of electricity generating power plants. *Energy, 52,* 210–221.
WNA. (2016). World Nuclear Performance Report 2016. World Nuclear Association, June 2016.

Author Biography

Dr. Ravi. B Grover occupies Homi Bhabha Chair instituted by the Department of Atomic Energy (DAE) and is a member of the Atomic Energy Commission. He studied mechanical engineering at Delhi College of Engineering, Delhi University, nuclear engineering at Bhabha Atomic Research Centre (BARC) and received a Ph.D. from the Indian Institute of Science, Bangalore.

He has been working in the DAE for about four and a half decades and was Principal Adviser during October 2010 to February 2013. As a nuclear engineer, he has worked in the areas of fuel and core thermal hydraulic analysis, process safety and plant dynamic analysis, process design of primary systems and equipments, and studies related to growth in energy requirements in India.

He conceptualized the setting up of the Homi Bhabha National Institute as a university level institute and concurrent with other responsibilities, he led it for about eleven years until 2016. He participated in negotiations with other countries and international agencies leading to opening of international civil nuclear trade with India.

He was conferred with a Padma Shri in 2014. He is a fellow of the Indian National Academy of Engineering, and the World Academy of Art and Science (Prof. Grover can be contacted at: rbgrover@hotmail.com).

Chapter 2
Changing Geopolitics of Energy Security and the Nuclear Power

Girijesh Pant

2.1 Introduction

The geopolitical coordinates of energy security are changing. Moving beyond the supply–demand security of hydrocarbon, the emerging energy security issues are going to be defined by global common concerns, be the UN declaration on Sustainable Energy for All or the pledge by more than two hundred countries to cut down carbon emission to limit the temperature increase to 1.5 °C above pre-industrial levels. Apparently in its revised terms and reference, energy security is embedded in a more complex ecosystem of transition which is expected to ensure availability, accessibility, affordability and acceptability on the trajectory of low-carbon energy system. It also calls for revisiting the issues such as the security for whom, security for which values and security against what threats (Cherp and Jewell 2014). In its extended define, it is also perceived as choice of energy system (Debaz 2016). This, however, could not have been conceived but for the innovation and transformation in energy technology and potentials and promises of the "digitalization of energy infrastructure" ensuring "greater energy security at a local, regional and national level (Molinaroli 2016)". The road map thus is being cartographed at global and local level; therefore, the future energy mix has to be correspondingly imagined not in isolation but in conjunction with each other. Surely, every country despite part of the global has to make its own choice of the mix. Apparently, the new energy mix is imagined in favour of clean and cleaner fuel. Nuclear is cleaner when compared with coal, oil and gas. It has to be the part of schema with larger profile. However, views are divided on role of nuclear in reaching the free carbon era. There are those who see nuclear in comparison with

G. Pant (✉)
School of International Studies, Jawaharlal Nehru University, New Delhi, India
e-mail: gpant2009@gmail.com

N. Janardhanan et al. (eds.), *Resurgence of Nuclear Power*,
https://doi.org/10.1007/978-981-10-5029-9_2

coal and oil and recommend it as substitute in power sector, while others see nuclear in terms of low-carbon bridge technology. It is argued that "Even if you want to be using 100% renewables, which not everyone does, it's going to take many decades to get there, probably sixty or seventy years at least. So, nuclear is an essential low-carbon bridge technology for those decades (Leggett et al. 2011)". It is interesting that in the context of climate change, nuclear has got new recognition. Under the umbrella of Nuclear for Climate initiative of French Nuclear Society, American Nuclear Society and European Nuclear Society, collective efforts are made to reposition nuclear power in the mainstream agenda of energy security as part of the solution (SFEN 2016).[1] This chapter aims to analyse the repositioning of nuclear energy not merely as supplementary but in the wider context of sustainable energy security. Nuclear is cleaner and has a role in mitigating environmental consequences of prevailing energy mix. It has a place in transition, but the choice in its favour is impacted by factors besides finance and the time lead, as will be elaborated in the following sections.

2.2 World Energy Market

The world energy market is witnessing transition and transformation simultaneously. It is much crowded, witnessing competition within the hydrocarbons and with clean energy. The geopolitics of energy is going to be played out at two levels. One on the terrain of substitution where the old energy players (hydrocarbon) would be desisting the entry of new players (energy alternatives) and within the domain of the hydrocarbon between old players, namely OPEC, and new suppliers such as USA, Canada, Australia and Russia and within energy alternatives themselves, e.g. trade war on local content among countries. This is further complicated by state policies shaping demand "by promoting consumption of certain fuels and restricting consumption of others (Makenzie 2015)". It is quite evident that unlike the past, the transition is going to be critically influenced by the intensity of competition among the different fuels. The intensity of competition, however, is going to be determined by the technological advances impacting on comparative costs and the regulatory mechanism by the governments. Ironically energy market never witnessed level playing field. Subsidies have been the source of perpetual market distortion (IEA 2014).[2] The present transition will also be witnessing a

[1]Nuclear for climate (N4C) is a grassroots initiative, which brings together scientists and professionals of the global nuclear community, and also citizens who believe that in order to fight climate change we have to act now. They believe that nuclear is part of the solution.

[2]Expresses concern about subsidies to fossil fuels, "which encourage wasteful consumption" and totalled $548 billion in 2013, over half of this for oil. Ten countries account for almost three-quarters of the world total for fossil fuel subsidies, five of them in Middle East (notably Iran and Saudi Arabia) or North Africa where much electricity is generated from oil and where nuclear power plants and even renewables would be competitive, but for those subsidies. The report

regime of subsidies. Importantly, transformational technology is enabling vital fuel substitution in energy end use. This has been most dramatic in power generation and distribution sector and is likely to be in transport sector as well. Equally, critical factor has been the efficiency improvement across the supply chain. Cumulatively, the impact on interfuel substation is more visible in volume and scale. It is argued that "As interconnectivity among energy markets grows, so too will the complexity confronting players across the energy value chain. Not only will the pace of change accelerate, but price shocks, regulatory changes, or technological breakthroughs in any given space will pose a much greater threat now than they used to as disruptions ripple across interconnected markets (Brognaux and Ward 2015)".

Trends indicate a more segmented market in terms of energy mix. No wonder it is referred in plural—Global Energy Markets (EIA 2015) Global Energy Markets. The OECD preference is for renewable and gas, and the emerging markets compulsions are to continue to rely on hydrocarbons specially coal along with renewable. The geography of energy consumption has changed and the geography of energy supply too. It is argued that "Trade patterns will evolve as commodity demand growth shifts to emerging markets. North America will disrupt traditional trade flows as it becomes an energy export province, with oil imports declining and eventually going into reverse. North America will also compete with Russia and the Caspian to become the world's foremost natural gas supplier (EIA 2015)". The geography of energy consumption is also moving with technological changes. Consequently, the geostrategic positioning of market is shifting. It is pivoting Asia though Yergin makes nuanced point that "We are experiencing a movement from the BRIC era to the Shale era, from a time of great demand and scarcity to a time of an abundance in supply and weaker demand".

Rise of clean energy despite the domination of hydrocarbons is the emerging reality of energy market. This is reflected in the investment flows and capacity addition. "In 2015, renewable energy set new records for investment and new capacity added. Investments reached nearly $286 billion, more than six times more than in 2004, and, for the first time, more than half of all added power generation capacity came from renewables (UNEP-Bloomberg 2016)". The graph below clearly shows that the power sector energy mix is going to be changed radically in favour of clean energy (Fig. 2.1).

(Footnote 2 continued)

advocates ensuring "that energy prices reflect their full economic value by introducing market pricing and removing price controls". Renewables subsides in 2013 are put at $121 billion and rising $45 billion of this being solar PV. Geographically, this is $69 billion for EU and $27 billion in USA. The report was unable to assign a figure for nuclear subsidies, which at present do not exist. The difficulty in reducing subsidies is discussed.

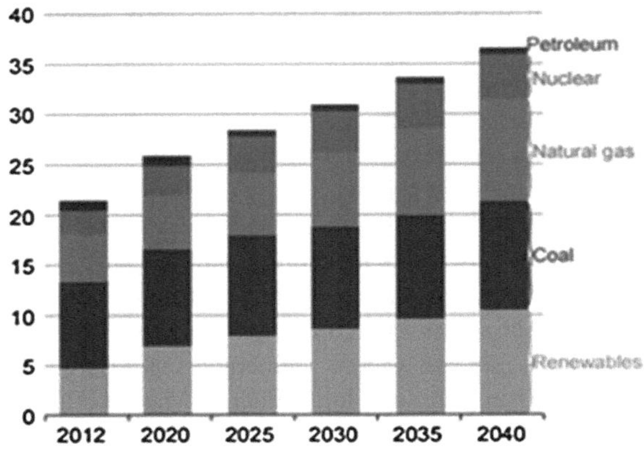

Fig. 2.1 World electricity generation, 1990–2040. *Source* US, EIA. International Energy Outlook 2016

2.3 Nuclear Energy in Global Power Market

Historically, it was on 26 June 1954 that the first time commercial use of energy from nuclear power plant commenced in Russiacs (ENS 2016) Russia. In the USA, President Eisenhower inaugurated the Shippingport Atomic Power Station on 26 May 1958. With growing recognition that nuclear is safe and cleaner energy, nuclear enjoyed wide acceptability. It was argued that in fifties and sixties, "The amount of energy produced by the fission of a single uranium atom is approximately 10 million times the energy produced by the combustion of a single coal atom (EBSCO 2016). The US Atomic Energy Commission celebrated the potential of nuclear by observing that it would become "too cheap to metre": the oil crisis of early seventies gave new impetus, and nuclear was seen as alternative source of power. In America under the Energy Independent Sentiment, President Nixon planned to build 1000 nuclear reactors by 2000 to augment domestic energy production. However, the planned expansion experienced setback with incidents such as Three Mile Island (1979) and Chernobyl (1986). The market for nuclear revived in 2000 euphemistically described as renaissance. "A parallel trend during the last two years has been a rebirth of serious interest in nuclear power. Economic, environmental, and political factors are now aligning to drive a 'renaissance' in the use of nuclear power for electrical energy production. Based on economic factors in electrical energy markets and the fact that the current fleet of nuclear plants is rapidly approaching the end of their initial licence periods, a strong economic incentive exists for new plant orders during the next 10–15 years (Wood et al. 2001)".

In 2011, Fukushima disaster came as a jolt, once again raising the issue of safety and security. The debate was once again triggered about the desirability of nuclear power. However, despite series of setbacks, the impact on the market has been of mixed nature. While countries such as Italy and Germany decided to phase out their nuclear plants, many others such as China, the United Arab Emirates, France, Poland, the UK and the USA decided to go with enhanced safety and security measures. The growth has been slow to stagnation during 1995–2014 when number power reactor went up from 436 to 437. However, the World Nuclear Association is optimist as it envisages the start-up of 266 new reactors by 2030 (Greens 2015).

IEA observes "nuclear is one of the world's largest sources of low-carbon energy (an average nuclear plant has the production of 4000 windmills), and as such, has made and should continue to make an important contribution to energy security and sustainability" (IEA 2016). It also acknowledges that nuclear power is the second-biggest source of low-carbon electricity worldwide after hydropower and that the use of nuclear energy has avoided the release of 56 billion tonnes of CO_2 since 1971, equivalent to almost two years of global emissions at current rates. According to World Nuclear Association (WNA 2015), nuclear power acquires importance in low carbonisation process, because it is environmentally benign and hence could play an important role in meeting the climate change target. It is estimated that the electricity demand which doubled from 1990 to 2011 is likely to grow from 19,004 to 34,454 TWh in the year 2011–2035. Importantly, it is Asia that the new surge is likely to take place averaging 4.0 or 3.6% per year, respectively, to 2035 (WNA 2015). The future dynamics of nuclear power indicates two critical shifts, one as mentioned above: Asian surge for nuclear power; second, the rise of nuclear in global power generation mix. This is substantiated by the World Energy Outlook 2015. Though the nuclear provides 11% of world electricity consumption, its share in OECD countries is 21%. Further, the global power (IEA 2015a, b) generation mix is shifting away from coal. Its share is likely to decline from 41% since 1990 to 30% in 2040. The shift is further enforced by the fact that the share of low-carbon technologies in total generation is to go up from one-third in 2013 to 47% in 2040 (IEA 2015a, b). The shift is also visible in installed nuclear capacities expected to grow from 543 Giga watt electrical (GWe) in 2030 to 624 GWe in 2040 out of a total of 10,700 GW. The increase is going to be concentrated in China (46%), India, Korea and Russia (30% of it together) and the USA (16%), but a decline of 10% in the EU. However, the rising share of nuclear power in the global power mix is to be only 12%, well below its historic peak (IEA 2014a, b, c)". According to IEA (2015a, b) most of the new nuclear plants are expected to be built in countries with price-regulated markets or where government-owned entities build, own and operate the plants, or where governments act to facilitate private investment (IEA 2015a, b)".

2.4 The Geopolitical Shift: Nuclear Energy and the Asian Pivot

The shift in economic gravity to Asia represented by the emerging economies such as China and India leads to corresponding rise in energy consumption in Asia. All estimates suggest that the future energy market is in Asia. It is also recognised that though hydrocarbon will remain central to Asian energy mix, the Paris Agreement puts obligation on them to diversify sources of energy and push for shift in favour of low-carbon energy. It is in the context of diversification of energy mix that nuclear has acquired importance in Asia despite Fukushima. Though Asia has been producing nuclear energy at small scale reflecting very low share in energy mix, the revival today is triggered by climate concerns. However, the drive for nuclear energy still remains constrained by the cost, safety and security more so of disposal of high waste. The leakage of radioactive tritium from one of America's oldest nuclear power plants—Indian Point Energy Center, owned by Entergy, just 25 miles north of New York City—levels 65,000% higher than normal has caused new alarm about the safety question (Shank 2016): Apparently the public opinion once again swinging against nuclear energy. USA had not constructed a new nuclear power plant in nearly 20 years when, in October of 2015, a plant in Tennessee was given green signal. However, as Table 2.1 shows despites apprehensions, countries are making choice for nuclear energy.

From geopolitical perspective, it is significant that the lead enjoyed by the USA is slipping away to China as it emerges with largest number of nuclear reactors. What is more critical that with the advantage of reverse engineering of Westinghouse AP 1000 reactor, "Chinese have come up with their own design that they soon hope to sell on the world market. Last month, the Chinese startled the world by signing a nuclear agreement with Great Britain whereby it will own 33.5% of the Hinkley Point reactor, Britain's first reactor in 20 years (Tucker 2015)". China is reportedly emerging as exporter of nuclear power plants. It has developed indigenously designed Hualong One reactor which was publicly demonstrated by China General Nuclear Power Group, to the business executives from Asian

Table 2.1 Nuclear reactors around the world

Country	Reactors operable	Reactors under construction	Reactors planned	Reactors proposed
USA	99	5	5	17
France	58	1	1	1
Japan	48	3	9	3
Russia	34	9	31	18
South Korea	23	5	8	0
China	22	27	64	123
India	21	6	22	35
Canada	19	0	2	3
UK	16	0	4	7
Ukraine	15	0	2	11
World Total	437	70	183	311

Source Richard Anderson, nuclear power: energy for the future or relic of the past? http://www.bbc.com/news/business-30919045

countries with the intent to export. “China’s push into nuclear power comes as many nations have been re-examining the risks of nuclear energy and its costs compared with natural gas and other fuels. Two dozen reactors are under construction across China today, representing more than one-third of all reactors being built globally, according to the International Atomic Energy Agency—The scale and pace of building has given CGN and other Chinese companies opportunities to bulk up on experience in the home market and gain skills in developing reactor parts, technologies and systems. That experience, combined with China’s lower costs of labour and capital, makes the new Chinese reactor potentially attractive to international customers, industry experts said (Spegele 2016)”. Chinese leap has made American nuclear industry to realise that market is slipping away. American neglect of the growing demand for nuclear energy in West Asia has paved way to others like Russia and China to pitch in with least competition.

Russian firm Rosatom is a leading player in nuclear energy market. The company is reportedly having contract worth $740 billion to build 30 reactors in 12 countries, including Turkey, Egypt, Saudi Arabia, Iran, India, Argentina, Bangladesh, Nigeria, Algeria and others. It is argued that both Russia and China enjoy the cost advantage and thereby will be the dominant player in nuclear reactor market. It is significant that though the two countries do compete, they have been collaborating as well. Rosatom is implementing a project in Jordan in collaboration with China. “In 2013, Rosatom got a contract for the construction of the first nuclear power plant in Jordan. Then, it was planned that the Russian side will finance half of the project cost of $10 billion. However, in the autumn of 2015, it was decided that China will take on about 50% of the costs. So far, Beijing has only funded Rosatom projects, which were carried out in China. As can be seen, fearing competition, Rosatom is committed as soon as possible to take a free niche in the market of nuclear technology, even if it means resorting to the help of its rival (Bokarev 2015)”.

The nuclear reactors beside source of energy have been the instrument of strategic leverage. Nuclear energy reactors have been used as disgusied step towards weaponisation. It is in this sense that it shifts the geopolitics towards new coordinates. This was acknowledged and lamented by Mr. Nakahara, Director of the Office for International Nuclear Energy Cooperation at METI, while speaking to the US–Japan Round Table, “the increased role that China and Russia are taking in the sphere, and expressed concern that the nuclear world has paid too little attention to the potential implications of such developments (Forumonenergy 2015)”. Further given the fact that timeline for constructing a power plant is quite long and the Russian contracts are largely on build-own-operate model ensures that the project gives additional diplomatic leverage to Russia (Armstrong 2015). It is estimated to be $1.2 trillion market (Fig. 2.2).

Nuclear energy has been instrument of US foreign policy. “American geopolitical, economic and technological leadership also predisposes the USA to lead and benefit from the energy sector transformation abroad. The global transition to a lower carbon economy plays directly to the comparative advantages of the USA: individual empowerment, innovation and engineering ingenuity (Koranyi 2016)”.

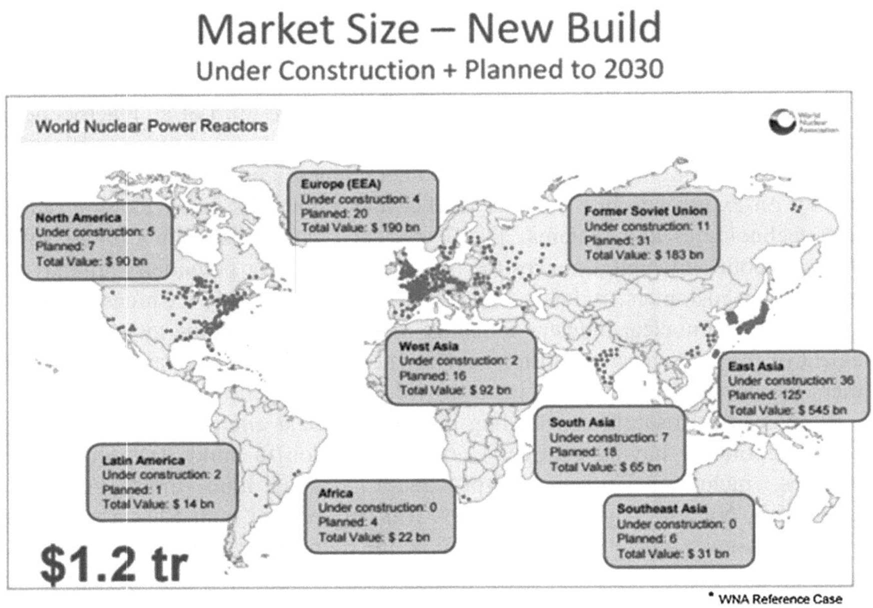

Fig. 2.2 Market size—new build. *Source* World Nuclear Association

Being leader in technology, USA has been in the market in a big way. Given the vast experience with civil nuclear energy with largest installed capacity, US nuclear industry has been top performing all along the value chain (Gottemoeller 2013). Twelve of the world's 25 highest-performing reactors are in the USA. Recognising the nature of competition, including financing, the state department has initiated a few initiatives like "Team USA" approach to civil nuclear engagement abroad to support industry. To strengthen support position of Director of Nuclear Energy Policy was created in January 2012. Similarly, "The Department of Commerce has established a Civil Nuclear Trade Initiative, the goal of which is to identify the US nuclear industry's trade policy challenges and commercial opportunities and coordinate public–private sector responses to support the growth of the US civil nuclear industry (Gottemoeller 2013)".

In the European theatre, nuclear energy debate has taken new twist with the cancellation of EPR[3] by Finish government on the grounds of delay and financial escalation. EPR under construction, at Flamanville (AREVA 2016) in France, is also seriously delayed and possibly in even deeper trouble because of concerns

[3]The EPR is a third-generation pressurised water reactor (PWR) design. It has been designed and developed mainly by Framatome (nowAreva) and Électricité de France (EDF) in France, and Siemens in Germany. In Europe, this reactor design was called European Pressurized Reactor, and the internationalised name was Evolutionary Power Reactor, but it is now simply named EPR (technical details: https://en.wikipedia.org).

about the quality of steel in the pressure vessel (Eco Watch 2015). "This is a serious blow to the pride of a country that is seen as the world leader in nuclear energy, with 75% of its electricity coming from 58 reactors (Eco Watch 2015)". It is reported that many plant are towards closer because "where markets are freer, it is harder for nuclear power operators to make money, and too risky for them to build plants from scratch" (The Economist 2015).[4] It is observed that in Europe, where market is deregulated, lower cost of electricity is making nuclear uneconomic (The Economist 2015). In Sweden, electricity prices at times have been below nuclear operating costs.

The economics of nuclear energy suggests that "Other things being equal, nuclear power's front-loaded cost structure is less attractive to a private investor in a liberalised market that values rapid returns than to a government that can consider the longer term, particularly in a regulated market that ensures attractive returns. Private investments in liberalised markets will also depend on the extent to which energy related external costs and benefits (e.g. air pollution, GHG emissions, waste and energy supply security) have been internalised. In contrast, government investors can incorporate such externalities directly into their decisions. Also important are regulatory risks and political support for nuclear power. All these factors vary across countries. In the Republic of Korea, the relatively high costs of alternative electricity sources benefit nuclear power's competitiveness. In China and India, rapidly growing demand for electricity encourages the development of all energy options (IEA 2015a, b)".

"The closures on both sides of the Atlantic represent a blow to an industry that for years talked about the prospect of a 'nuclear renaissance' based on the merits of large-scale, low-carbon energy. They show how Western governments have waffled in their support for the technology, opting to subsidise renewables rather than putting a real price on carbon emissions that would more heavily penalise dirtier fuels, such as coal and gas, and thereby promote nuclear energy. The closures also highlight how green NGOs, many of which were created to oppose both military and civilian use of atomic energy, have influenced the debate on nuclear power. Many greens still see it as an evil akin to global warming (IEA 2015a, b)" (Fig. 2.3).

From geopolitical perspective, it is significant that while West is having second thought about the desirability of nuclear reactors despite its being source of clean energy, Asia is going big way despite Japanese experience. Interestingly out of six

---

[4]Exelon, based in Chicago and the largest operator of nuclear plants in America, says that five of its 14 plants are vulnerable because of economic factors, including Three Mile Island's Unit One, which it owns. "It's ironic. People ask why we still operate a reactor there. But if gas prices were not [so low], it would be making money", says David Brown of Exelon.

On October 13th Entergy, Exelon's rival, which is based in New Orleans, said it would close its Pilgrim nuclear plant in Massachusetts, partly because its costs, at about $50 a megawatt-hour (MWh), are higher than electricity prices in the state, which have fallen to about $45/MWh. As The Economist went to press, it was due to decide whether to close a third, Fitzpatrick, in New York State. In December, it closed one in Vermont, the fourth American nuclear plant to shut in the past two years.

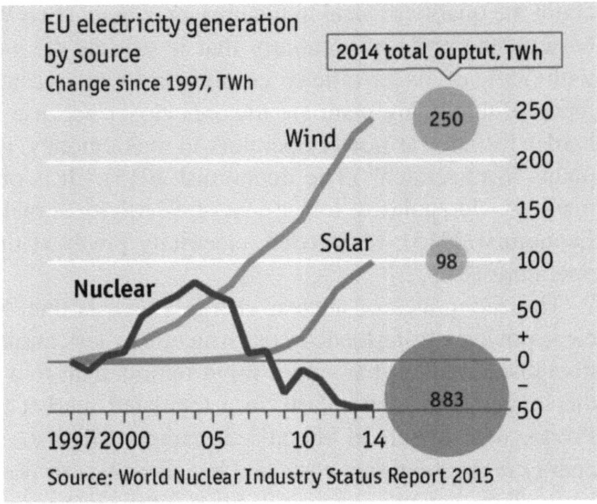

Fig. 2.3 EU electricity generation by source. *Source* World Nuclear Industry Status Report, 2015

Table 2.2 Nuclear power in Asia and involvement with the nuclear fuel cycle

	Power reactors operable or in operation	Power reactors under construction	Power reactors planned	Research reactors
Australia				1
Bangladesh			2	1
China	30	24	40	16
India	21	6	22	4
Indonesia			1	3
Japan	43	3	9	14
S. Korea	25	3	8	2
N. Korea			0	1
Malaysia			0	1
Pakistan	3	2	2	1
Philippines			0	1
Thailand			0	1 + 1
Vietnam			4	1
Total[b]	128	40	89	49[a]

Source WNA (2015)
[a]48 research reactors operable, 1 under construction
[b]The total includes six reactors in operation, plus two under construction, on Taiwan. It also has one research reactor. Taiwan has no other stages of the fuel cycle

countries trading nuclear technology, four belong to Asia: France (AREVA), USA (General Electric, Westinghouse), Russia (ROSATOM), Japan (Toshiba, Hitachi), South Korea (KEPCO), China (CGNPC). The high finance and long timeline are other features of this option (Table 2.2).

2.5 Repositioning Nuclear Power in the Energy Security Discourse

The climate change agenda has brought sustainability central to energy security discourse. The choices are being made in comparative perspective of that energy mix which contributes in accomplishing the goal set by the Paris conference in December 2015. The dirt of the radioactivity emission of nuclear energy is fading in favour of its decarbonised nature. It is rightly observed that "When you weigh the clean energy benefits against the dangers of air pollution from other forms of energy, it's clear that nuclear energy is punching far above its weight class (Whitman 2016)". Thus, countries are repositioning the need for nuclear reactor to meet their respective commitments with calculated risk.

President Obama has in his Climate Action Plan reiterated the positive role of nuclear energy towards clean energy transition and promised helping hand to struggling nuclear industry in USA. "The President's FY 2016 Budget includes more than $900 million for the Department of Energy (DOE) to support the US civilian nuclear energy sector by leading federal research, development and demonstration efforts in nuclear energy technologies, ranging from power generation, safety, hybrid energy systems and security technologies, among other things. DOE also supports the deployment of these technologies with $12.5 billion in remaining loan guarantee authority for advanced nuclear projects through Title 17. DOE's investments in nuclear energy help secure the three strategic objectives that are foundational to our nation's energy system: energy security, economic competitiveness and environmental responsibility (US Embassy 2015)".

In the context of geopolitical imperatives of nuclear energy, it is argued that (IEA 2014a, b, c)[5] "Uranium resources are spread across five continents and are available to satisfy the needs of the global economy in the twenty-first century. However, geological availability of an energy source is not enough to guarantee the security of energy supplies: unpredictable interruptions of extraction and transport and a high level of uncertainty about future supplies due to the high market power of exporting countries can negatively affect the expectations of consumers about future access" (IEA 2014a, b, c) (see footnote 5). It is clear from above that as low-carbon technology, nuclear reactor needs to be factored in transition regime, but apart from cost, waste disposal and proliferation, it may be underlined that it is composed of many components small and big which requires a support of dynamic, robust and diversified global supply chain (NEI 2016). Apparently sourcing involves large number of companies and workers, e.g. "the current Areva project in Finland has more than 4000 employees on site from 55 different countries and their new French build has around 100 suppliers from the global supply chain (Hoggett 2014)". As shown in the

[5]There is little or no likelihood of any uranium-producing country or region gaining a monopoly. Uranium resources are distributed evenly, and 35% of global uranium resources are located in OECD countries. Australia alone holds 23% of global resources, and around one-quarter of global resources are located in Eurasia, alongside significant resources in Africa and Latin America.

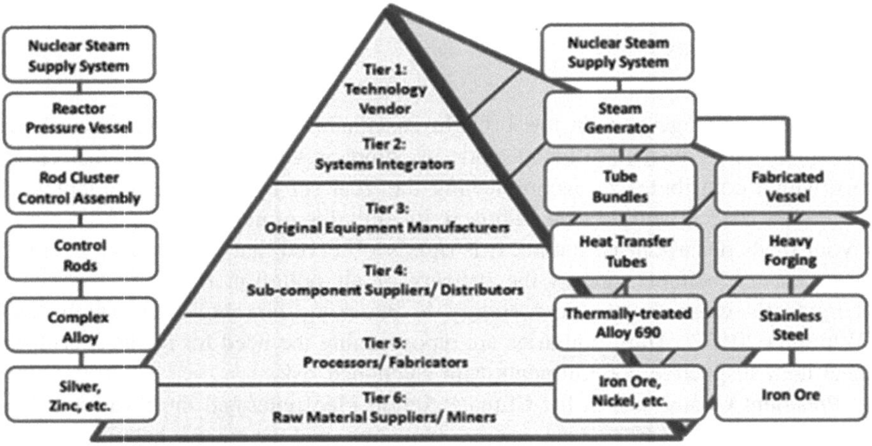

Fig. 2.4 World nuclear supply chain—an overview. *Source* World Nuclear Association, March 2014

Fig. 2.4, the supply chain is in pyramid shape with six tiers. "The top tier comprises the Technology Vendor, who are the main contractors for a plant. There are currently nine consolidated vendors operating in the global market, with four dominating the market (AREVA, Hitachi-GE, Toshiba Power Systems and Mitsubishi Heavy Industry). Tier two is of—System Integrators (e.g. reactor pressure vessel and steam generators); Tier three—Original Equipment Manufacturers (e.g. rod cluster control assembly); Tier four—Sub-component Suppliers/Distributors (e.g. control rods and heavy forgings); Tier five—Processors/Fabricators (e.g. alloys); and Tier six—Raw Material Suppliers/Miners (e.g. silver, zinc) (Hoggett 2014)". Each component in the supply chain involves investment and specialised skill; hence, there have been very few players. There are only four companies globally capable of producing ultra-large forging, operating in Japan, China and Russia.

The complexity of nuclear power plan clearly brings out that in the context of Paris Agreement commitment, the supply uncertainties could pose problem. "From a supply chain perspective, risks appear to increase if a supply chain is reliant on a limited number of companies, technologies or markets, whereas resilience increases if the number of companies, networks, connections, etc., is large, as this creates alternative options for bringing forward a low-carbon technology at an affordable cost (Hoggett 2014)".

Large technologies offer less resilience than smaller-scale technologies. Innovation also appears to be slower, with long life cycles between generations of plant, and once deployed, they remain on the system for decades, potentially constraining the development of the system and other technologies. Smaller-scale technologies show quick rates on innovation and can be quickly deployed and improved. Arguably, then, from an energy security and low-carbon transition perspective, there is something inherently more secure about smaller-scale technologies (Hoggett 2014).

From the above, it is clear that nuclear energy despite being clean can play to its full potential only if it also like other energy sectors, namely renewables (solar power), extraction (hydraulic fracturing), storage (advanced batteries) and consumer efficiency (advanced thermostats), is engaged with technological innovation (Brinton 2015). It is argued that "without significant advances in nuclear reactor and fuel cycle technologies—advances yielding cost reductions, shorter cycle times, a greater focus on passive safety, and other improvements—nuclear is unlikely to play that role (Yurman 2016; Lester 2016)". The road map suggested is as follows (Energy Collective 2016):

- First—extend the operational lifetime of the existing fleet. Innovation focuses on cost control and efficient operation. It covers the current era to the end of the 2030s.
- Second—build a new, expanded fleet, primarily of large and small light-water reactors, and bring to commercial deployment advanced nuclear technologies for use in power generation, but also desalinization, process heat and production of fuels for the transportation sector. It begins in the 2030s and extends to the end of this century.
- Third—develop a second generation of advanced nuclear technologies in the post-2050 timeframe to broaden their use globally.

Further, the possibilities of small modular reactor[6] have enhanced the prospecting of nuclear energy. It is a widely agreed assessment that SMRs are going to be the next-generation reactor. These SMR's are likely to be sized between 10 and 300 MW. These SMRs being small in size "are designed to have many components fabricated and assembled offsite, thus reducing the time and complexity of plant construction and increasing potential plant locations. SMR designs generally have their reactors buried in the ground away from weather hazards and are often designed to use passive cooling systems that are not vulnerable to power outages, further increasing the safety of the plan (BRIGGS 2016)".

They generally have their reactors buried in the ground, away from weather hazards. These designs often use passive cooling systems that are not vulnerable to power outages, increasing the safety of the plant (BRIGGS 2016). Countries are investing on designs to gain the advantage of market. Brittan has planned to spend pound 250 million", "position the UK as a global leader in innovative nuclear technologies" and pave the way "towards building one of the world's first SMRs in the UK in the 2020s". There is no shortage of contenders, with companies from the

[6]Small Modular Reactors (SMRs) are nuclear power plants that are smaller in size (300 MWe or less) than current generation base load plants (1000 MWe or higher). These smaller, compact designs are factory-fabricated reactors that can be transported by truck or rail to a nuclear power site. SMRs will play an important role in addressing the energy security, economic and climate goals of the USA if they can be commercially deployed within the next decade" (Refer: https://www.energy.gov).

USA to China and Poland all wooing the UK with their proposal (Carrington 2015)". Commercialisation of SMR opens huge and variety of possibilities having distinct geopolitical narrative. The idea of floating power plant is one such possibility where "an SMR could be put on a barge, taken to a country, plugged into grid from the port and then, when its fuel was used up, sail back again. 'For newcomer countries [to nuclear], that could be a very attractive way to do it', she said. It has happened before: in the late 1960s a former US military nuclear-powered ship moored by the Panama canal and provided onshore electricity (Carrington 2015)". Russia has been planning a fleet of floating and submersible nuclear power stations to exploit Arcticoil and gas reserves. Nuclear is returning, but it has to evolve through technological transformation to become cheaper, safer, more scalable successors to the big nuclear plants of old to meet the parameter of energy security.

The geopolitics of energy security defined by control over of hydrocarbons overtly or covertly is being redefined by the structural changes in energy sector revolutionised by technological changes and the existential necessity to move towards low-carbon society. Consequently, low-carbon fuel is going to be the preferred source of energy security. In the context of shift in global geopolitical parameters driven by technological changes with high disruptive impact, the virtual realities are going to become the key factor of geopolitical influence and power relations. In the energy sector, the transition to low-carbon fuel is going to be the central of defining power relation between producer and consumer because hydrocarbon market is witnessing demand peak. The move to low-carbon energy regime critically hinges upon the pace of technological transformation along with institutional changes. Thus, it is the low-carbon technology that is going to be playing out the geopolitics of low-carbon fuel energy. It is righty argued that those who are investing on new technology would be the leaders in the low-carbon energy market. Surely, the geopolitics of low-carbon energy security is going to witness a race "to occupy and monopolise the low-carbon technology as well as its product market expresses the current new connotation of superpower's geopolitical strategy; that is, who dominates low-carbon technology would occupy the high ground (Wang et al. 2012)". In the low-carbon technology era, as argued above, nuclear has its own role towards low-carbon energy security, but as the EU document on The Sustainable Nuclear Energy Platform 2015 argues "The long-term sustainability sustainability of nuclear energy will be ensured by Gen IV fast neutron reactors and closing the fuel cycle, minimising the nuclear waste and offering a transmutation option as well. This will require a large R&D programme for supporting the construction of reactors (prototypes, research facilities, demonstrators) and related fuel cycle facilities (SNETP 2015)". It is innovation and technological breakthroughs: the next-generation low-carbon technologies technologies that are going to cartograph the geopolitical frontiers of energy security.

References

AREVA. (2016). *Flamanville 3-France.* Available at: http://www.areva.com/EN/operations-2397/france-flamanville-3.html. Accessed September 30, 2016.

Armstrong, I. (2015). *Russia is creating a global nuclear power empire.* Available at: http://globalriskinsights.com/2015/10/russia-is-creating-a-global-nuclear-power-empire/. Accessed May 1, 2016.

Bokarev, D. (2015). *New eastern outlook.* Available at: http://journal-neo.org/2015/11/24/is-there-a-peaceful-nuclear-race-between-russia-in-china-in-the-middle-east/. Accessed May 1, 2016.

BRIGGS, L. (2016). *Advanced energy perspectives.* Available at: http://blog.aee.net/news-tomorrows-technology-today-plus-nothing-mickey-mouse-about-this-solar-installation. Accessed April 18, 2016.

Brinton, S. (2015). *The advanced nuclear industry.* Available at: http://www.thirdway.org/report/the-advanced-nuclear-industry. Accessed June 1, 2016.

Brognaux, C., & Ward, N. (2015). *BGC perspectives.* Available at: https://www.bcgperspectives.com/content/articles/energy-environment-when-fuels-compete-evolving-dynamic-global-energy-markets/. Accessed March 30, 2016.

Carrington, D. (2015). *Guardian.* Available at: https://www.theguardian.com/environment/2015/nov/24/mini-nuclear-reactors-answer-to-climate-change-crisis. Accessed June 1, 2016.

Cherp, A., & Jewell, J. (2014). The concept of energy security: Beyond the four as. *Energy Policy, 75,* 415–421.

Debaz, J. (2016). *The rise of energy citizen.* Available at: http://www.kib.be/articles/1185/the-rise-of-energy-citizens. Accessed December 5, 2016.

EBSCO. (2016). *EBSCO host.* Available at: http://connection.ebscohost.com/science/nuclear-power/history-nuclear-energy-production. Accessed December 5, 2016.

Eco Watch. (2015). *Nuclear giants take a huge hit.* Available at: http://www.ecowatch.com/nuclear-giants-take-a-huge-hit-1882045728.html. Accessed June 1, 2016.

EIA. (2015). *Energy information administration.* Available at: http://www.eia.gov/pressroom/presentations/sieminski_09212015.pdf. Accessed February 22, 2016].

Energy Collective. (2016). *Why the next generation of nuclear energy technology innovation won't be like the last one.* Available at: http://www.theenergycollective.com/dan-yurman/2308648/why-nuclear-industry-must-respond-fourth-industrial-revolution. Accessed December 3, 2016.

ENS. (2016). *European nuclear society.* Available at: http://www.euronuclear.org/. Accessed December 5, 2016.

Forum on Energy. (2015). *Forum on energy.* Available at: http://forumonenergy.com/2015/11/06/the-shifting-geopolitics-of-nuclear-energy/. Accessed June 1, 2016.

Gottemoeller, R. (2013). *US department of state.* Available at: https://2009-2017.state.gov/t/us/209768.htm. Accessed January 2, 2017.

Greens, J. (2015). *The ecologist.* Available at: http://www.theecologist.org/News/news_analysis/2732640/running_in_reverse_the_worlds_nuclear_power_renaissance.html. Accessed May 1, 2016.

Hoggett, R. (2014). Technology scale and supply chains in a secure, affordable and low carbon energy transition. *Applied Energy, 123,* 296–306.

IEA. (2014a). *Climate and nuclear power.* Paris: International Energy Agency.

IEA. (2014b). *World energy outlook.* Paris: International Energy Agency.

IEA. (2014c). *World energy outlook.* Available at: http://www.worldenergyoutlook.org/weo2014/. Accessed December 5, 2016.

IEA. (2015a). *Climate change and nuclear energy.* Paris: International Energy Agency.

IEA. (2015b). *World Energy outlook.* Paris: International Energy Agency.

IEA. (2016). *Nuclear Energy.* Available at: https://www.iea.org/about/faqs/nuclear/. Accessed May 1, 2016.

Koranyi, D. (2016). *Atlantic council.* Available at: http://www.atlanticcouncil.org/publications/reports/a-us-strategy-for-sustainable-energy-security. Accessed June 1, 2016.

Leggett, J., Stevens, P., Thomas, S., & Tindale, S. (2011). *Chatham house debate: Is nuclear a sustainable energy option?*. London: Chatham House.

Lester, R. K. (2016). A roadmap for US nuclear energy innovation. *Issues in Science and Technology, 32*(2), 7.

Makenzie, W. (2015). *Global energy markets set for structural shift by 2035*. Available at: https://www.woodmac.com/analysis/energy-view-2035. Accessed April 2, 2016.

Molinaroli, A. (2016). *World economic forum*. Available at: https://www.weforum.org/agenda/2016/03/perspective-distributed-digital-and-demand-side-energy-technology-implications-for-energy-security/. Accessed April 2, 2016.

NEI. (2016). *Exports & trade*. Available at: https://www.nei.org/Issues-Policy/Exports-Trade. Accessed December 3, 2016.

SFEN. (2016). *Nuclear for climate*. Available at: http://www.sfen.org/en/nuclear-for-climate. Accessed December 5, 2016.

Shank, M. (2016). *Proof nuclear energy is huge security risk*. Available at: http://fortune.com/2016/03/08/nuclear-energy-security-risk/. Accessed April 1, 2016.

SNETP. (2015). *Sustainable nuclear energy technology platform*. Available at: http://www.snetp.eu/wp-content/uploads/2016/01/SNETP-DEPLOYMENT-STRATEGY-2015-WEB.pdf. Accessed June 1, 2016.

Spegele, B. (2016). *China Inc.'s nuclear-power push*. Available at: https://www.wsj.com/articles/china-inc-s-nuclear-power-push-1456251331. Accessed May 1, 2016.

The Economist. (2015). *Future of nuclear energy: Half-death*. Available at: http://www.economist.com/news/international/21677243-nuclear-power-emits-no-greenhouse-gases-yet-it-struggling-rich-world-half-death. Accessed May 15, 2016.

Tucker, W. (2015). *Russia and China are becoming nuclear titans*. Available at: http://www.realclearenergy.org/articles/2015/11/05/russia_and_china_are_becoming_nuclear_titans_108868.html. Accessed May 1, 2016.

U.S. Embassy. (2015). *US embassy and consulates in Brazil*. Available at: https://br.usembassy.gov/fact-sheet-obama-administration-announces-actions-clean-energy-strategy/. Accessed December 2, 2016.

UNEP-Bloomberg. (2016). *Global trends in renewable energy investment 2016*, Frankfurt: Frankfurt School-UNEP Centre/BNEF.

Wang, L., Gu, M., & Li, H. (2012). Influence path and effect of climate change on geopolitical pattern. *Journal of Geographical Sciences, 22*(6), 1117–1130.

Whitman, C. T. (2016). *Yahoo finance*. Available at: http://finance.yahoo.com/news/why-nuclear-energy-help-fight-180008048.html. Accessed June 2, 2016.

WNA. (2015). *World energy needs and nuclear power*. Available at: http://www.world-nuclear.org/information-library/current-and-future-generation/world-energy-needs-and-nuclear-power.aspx. Accessed January 5, 2016.

Wood, T., Johnson, W., & Parker, B. (2001). *Economic globalization and a nuclear renaissance*. Richland: Pacific Northwest National Laboratory.

Yurman, D. (2016). *Neutron bytes*. Available at: https://neutronbytes.com/2016/01/01/why-the-nuclear-industry-must-respond-to-the-fourth-industrial-revolution/. Accessed June 5, 2016.

Author Biography

Prof. Girijesh Pant is Professor of Energy Studies at the School of International Studies where formerly he was the Dean. His specialisation includes Development Studies, Political Economy of West Asia, The Gulf Economies, Global Energy Affairs and India's Energy Security. He has written extensively on these subjects in national and international platforms. He has been Vice Chancellor of Doon University, Dehradun, and GGD University, Bilaspur (Prof. Girijesh can be contacted at: gpant2009@gmail.com).

Chapter 3
The Expectant Global Nuclear Energy Renaissance: Movers, Shakers and Spoilers

A. Vinod Kumar

3.1 Introduction

The tsunami that struck the coast of Fukushima in 2011 was a milestone event that not just transformed the shape and character of the global nuclear industry, but also resulted in the shifting of its growth impetus from the Western industrial hubs to the emerging economies. A handful of trends could be listed for this transformation. First, a wave of public protests had erupted around the globe against nuclear power over safety concerns, which has affected the prospects of many planned nuclear projects and also prompted some industrially advanced states, especially in Europe, to reconsider nuclear energy as a sustainable energy source besides contemplating shutdown of older plants. Second, emerging powers in the developing world—notably China and India—are increasingly placing their faith on nuclear power as a panacea to address their energy demands. Though resistance from anti-nuclear groups might be threatening to delay, and in some cases derail, the nuclear power projects in these nations, the expansion of the nuclear energy sector in these markets is also being shaped by a new architecture of rules, guidelines and structures that are intended to create a safe, an accountable and a vibrant sector.

It is widely perceived that the turnaround of the global nuclear power will be spearheaded by the massive expansion of nuclear energy sector, especially in the Asian region. Does this imply the advent of a renaissance in the nuclear energy sector, and if so what does it mean for this sector? Will the emerging economies in the region overcome the numerous challenges and harness its opportunities for the renaissance to take shape? Who will be the key players driving this growth, revival and rejuvenation story, and what are drivers and impediments to its fulfilment? This paper undertakes a reality check of this burgeoning ecosystem and examines whether a nuclear energy renaissance is actually in the offing in the Asian region?

A. Vinod Kumar (✉)
Institute for Defence Studies and Analyses, New Delhi, India
e-mail: vinujnu@gmail.com

N. Janardhanan et al. (eds.), *Resurgence of Nuclear Power*,
https://doi.org/10.1007/978-981-10-5029-9_3

3.2 Why a 'Renaissance'?

The term 'renaissance'—analogised from the social–cultural revolution in Europe between the fourteenth and seventeenth centuries—is generically referred to the transition of eras, paradigmatic shifts or other transformations of colossal nature, which could change a way of life or character of a system. In the nuclear energy sector, renaissance has different connotations. Most descriptions relate the widespread anticipation of a global revival of the nuclear energy sector at the dawn of the millennium to the drive for clean fuels as replacement for fossil fuels, prompted by concerns of climate change. For the many industrially advanced states with a significant nuclear energy base, renaissance denotes the promise of rejuvenation of the sector after a prolonged phase of slowdown in new nuclear builds as also its economic revival through expansion. Nuclear renaissance for the emerging economies, however, is about providing the financial and technological wherewithal to facilitate growth of their nuclear energy infrastructure, from their current nascent stages, and expand nuclear energy as a source to meet their fast-growing electricity generation and economic needs.

According to the World Nuclear Association (WNA), 'the share of nuclear in world electricity has showed decline from about 17 to 11.5% since the mid-1980s, though output from nuclear reactors actually increased, albeit not enough to match the growth in global electricity consumption (WNA 2015a)'. Various factors have contributed to this downslide since the 1970s, which stretched all the way for over three decades, and have only showed intermittent signs of revival, never a convincing turnaround. Two major nuclear accidents—at the Three Mile Island plant in 1979 and the Chernobyl nuclear facility in 1986—had a far-reaching impact on the global public attitude towards nuclear energy and resulted in a perennial fear about their safety from radioactive fallouts. This factor had largely driven the investments in renewable energy resources, including solar and wind, though their ability to emerge as a sustainable power source has always been constrained in terms of quantity and large-scale distribution, despite quantum technological strides in these segments.

The continuing dependence on traditional fossil fuels and their spiralling effect on the environment and climatic patterns, however, have meant that the lasting quest for a clean as well as a source for large-scale production of electricity will continue to generate interest, and long-term dependence, on nuclear energy. This ought to happen despite the persistent safety concerns which could likely increase the costs of constructing large nuclear plants, and besides the sustained investments, the industry must commit to the long-term development of new-generation and advanced reactor technologies that could be safer, proliferation-resistant and economic. The prospects of a nuclear energy renaissance therefore is to be determined by how nations will strike a balance between these push and pull factors in facilitating the development of a robust nuclear industrial base. This will be crucial not just to realise the promise of a nuclear renaissance but also to shape a vibrant global nuclear energy industry.

3.2.1 The Struggles in the US Nuclear Industry

The catalyst for a nuclear renaissance has a lot to do with the way the nuclear energy sector in the USA, with highest number of operational reactors (WNA 2016a), has evolved in the last few decades. Despite having close to a hundred nuclear power reactors, and accounting for over 30% of electricity produced by nuclear worldwide, the USA had experienced more than three decades of stagnancy in this segment (WNA 2016b). After the last contract for a nuclear plant coming in 1973, it took over 30 years for the next application to come up before the US Nuclear Regulatory Commission in 2007 seeking licence for a nuclear power plant (NEI 2016). The hiatus of three decades has been attributed to a variety of reasons, though the all-pervading concerns and paranoia over nuclear safety have been embedded in the American psyche after the Three Mile Island and Chernobyl. The renewed interest since the first decade of the millennium was largely riding on the hopes of a renaissance following the calls by environment groups in favour of nuclear energy as a clean fuel alternative (Barrett 2015). With global dialogue and initiatives to tackle climate change gaining momentum in those early years of the millennium, it was widely hoped that an impending rejuvenation of nuclear energy sector in the USA would spearhead a global revival of this industry. The results, though, were not as anticipated.

In the early years of this revival phase, around 17 companies had applied or had plans to seek licences from the US Nuclear Regulatory Commission (NRC) for over 30 new reactors. However, currently only four reactors are under construction even as ten other existing nuclear plants were slated for phased closure owing to high operational costs, combined with lower power prices and regulatory issues, among others.[1] In the meantime, the USA witnessed a premature shutdown of eight reactors leading to a decrease of around 5700 MW capacity in the last two decades with the addition to the grid being only around 3470 MW (WNA 2016b).[2] This picture of fluctuating trends explains why the promise of a nuclear energy renaissance remained unfulfilled even in established nuclear industrial bases like the USA. While on the one hand, the push for clean fuel and low-carbon energy initiatives has buttressed opportunities for nuclear power, on the other hand, a set of countervailing forces such as the fall in gas prices, greater economic viability of renewable energy sources and nuclear safety concerns have stymied nuclear energy from emerging to the centre stage of energy planning.

---

[1]The units under construction include Vogtle 3 and 4 in Georgia and V.C. Summer 2 and 3 in South Carolina.

[2]The two reactors that came online were Comanche Peak 2 and Watts Bar Unit 1 along with restarting of the Browns Ferry Unit 1. Further, Watts Bar Unit 2 received license in October 2015.

3.2.2 Revive or Phase-Out: Europe's Dilemma

In fact, this has been a common phenomenon across the developed world, and especially in Europe, where many countries, including Germany, Italy, Switzerland, Belgium and Spain, have announced plans to phase out nuclear from their energy mix (EnergyTransition 2015). Of them, Germany's decision to shift to low-carbon and environmentally sound energy sources (*Energiewende*) was intensely debated as it came after public angst following the Fukushima incident, but not without questions on whether the largest economy in Europe could power its industries and society without large-scale power plants, and whether renewable sources could suffice in this mission. Interestingly, from around 29% in 2000, Germany has cut down the share of nuclear power to around 15% by 2014 and plans total shutdown by 2022 (CEW 2015). There are, however, notable exceptions in Europe like France, which produces nearly 75% of its electricity from nuclear power and deems nuclear energy as central to its energy security as well as clean energy vision (WNA 2016c).

On the other hand, despite the persuasive movements across Europe of a nuclear-free low-carbon energy sector, the region continues to throw a confusing picture of sustainability. This is an outcome of gas price fluctuations, unreliability of renewable fuel sources during winter months and volatility in the electricity prices affecting both consumers and viability of utility operators—all reviving debates and interest on nuclear as a sustainable and long-term clean energy option (EnergyPost 2016; Meyer et al. 2016). However, sustenance of nuclear projects in some European states may now look like an interim measure (or bridging technology) (Tindale 2011), as the trend across Europe is to enhance output from or prepare for overall transition to renewable sources. Yet, numerous hurdles for the renewable sector such as land occupancy, weather dependency as well as comparatively lower generation capacities in terms of power per plant will ensure that nuclear remains in contention for the long haul.

3.3 The Expectant Renaissance in Nuclear Energy

What are the trends that indicate the prospects of a renaissance happening, especially since earlier signs, like in the USA, did not end in fruitful outcomes? A handful of global indicators could be highlighted to support the assumption that a renaissance might be in the offing in the global nuclear industry, though with inherent challenges. First, the hopes of revival in many established nuclear bases such as USA and Russia even as some European nations such as UK, France and the Scandinavians have reposed faith in nuclear power, which also indicates a revisiting of phase-out policies in the industrial world. Second, the massive confidence that the developing world has placed on nuclear energy as evident from the numerous expansion plans of those nations with the existing nuclear energy infrastructure, as also the decision of many small and emerging economies to make their plunge into nuclear energy. Third is a host of systematic factors that has the potential to drive the renaissance, including the galloping pace of reactor technology development, best practices emerging in civil nuclear liability laws, increasing access of insurance and risk management tools, and above all the imperative of nations to promote clean fuel to meet climate change mitigation targets.

3.3.1 Industrial World Still Relies on Nuclear Energy

While the US revival efforts have been ongoing since the beginning of the millennium despite numerous constraints, Russia is on a faster pace in expanding its nuclear energy infrastructure with notable advances made in launching its Generation-III fast neutron reactors as well as in ensuring its technological footprint in emerging markets through exports of its VVER light water reactors (NewEurope 2016). Besides its existing tally of 36 operating reactors with a total capacity of 27,000 MW, Russia plans to add one large reactor per year till 2028, with around 22 planned units running up to around 21,000 MW capacity, alongside an equal number earmarked for exports (WNA 2016f).

As for Europe, where feasibility of nuclear power continues to remain in popular debate, countries that have preferred to retain nuclear power are announcing new projects or seeking to replace ageing ones with new units. The UK, with plans to retire half of 15 operational reactors by middle of next decade, has opted for a foreign-funded consortium model to kick off its Hinkley Point nuclear project, with Chinese and French state-run energy firms as collaborators (BBC 2016; Riley and Mullen 2016). This is beside an intended plan to allow the China General Nuclear Corporation to build a nuclear plant at Bradwell, as agreed during the Chinese President's visit in October 2015. France, which is heavily nuclear energy-dependent, is seemingly weighing on Areva's Generation-III European Pressurised Reactors (EPR) not just to display its continuing reliance on nuclear energy but also to establish its international standing as a nuclear supplier and technology incubator.

Despite reported timeline delays and cost outruns (Guardian 2016), the EPRs at Flamanville plant in France, Olkiluoto-3 in Finland as well as the proposed project at UK's Hinkley symbolise a renewed thrust across Europe for nuclear power, as also the role played by other new-generation reactors in this renewal. On the other hand, the actual relevance of the nuclear industry in Europe is in the comparatively nascent economies in Scandinavia, especially Finland and Sweden, which have preferred to retain nuclear as a power source and probably delay or avoid a total transition to renewable means. Along with unit 5, which is currently under construction, Finland is planning further expansion for units 6 and 7 in Olkiluoto (WNA 2016d). Another project, run by Fennovoima, is under construction to set up a VVER-1200 reactor and has a unique model of co-ownership by Rusatom Overseas (Milne 2016b).

Sweden, on the other hand, produces 40% of its electricity from nine existing reactors and plans to replace the ageing ones with an ambitious plan of 10 new reactors (Milne 2016a). However, Sweden's contribution to the nuclear revival is more significant. The nation not just revised its earlier plan, finalised in 1980s, to close nuclear and forego life extension of reactors, but removed a tax from 2017 that discriminates against nuclear power while subsides wind and biomass to progress towards the goal to total renewable energy by 2040. A certain fillip to nuclear energy, this reversal also means that stringent anti-nuclear sentiments that prevailed in Scandinavia following the Fukushima incident, and phase-out models like Germany's *Energiewende*, has given way to more realistic approaches towards nuclear energy. In Europe, these revival signs could be signs of an impending renaissance as it ignites fresh life into the industry.

3.3.2 Developing World as the Catalyst

At the other end of the spectrum, an expectant renaissance in nuclear energy is all about 'a great leap forward' for many countries in the developing world for whom access to sustainable means of energy is closely linked to the economic progress and upliftment of their societies. Nuclear energy has traditionally been an elitist preserve with the developed and the industrial world always controlling the technology and restricting access through non-proliferation structures including the Treaty on the Non-Proliferation of Nuclear Weapons (NPT) and supplier cartels like the Nuclear Suppliers Group (NSG). Even those countries in underdeveloped regions with major uranium deposits, like in some African countries, could not harness the economic potential of their natural resources owing to the domination and control of advanced nations over the affairs of the atom, and its normative structures.

Consequently, very few countries in the third world have been successful in establishing a nuclear industry that could be on par with their peers in the developed world. Even those who managed to set up a comparatively active base—such as India, China or Brazil—had perennial struggles with the development of reactor and reprocessing technologies, access to fissile materials or in dealing with denial regimes over attempts to develop indigenous capabilities or gaining strategic autonomy that mismatched with the global norms perpetuated by the Western-oriented liberal security community. The renaissance, therefore, for the developing world is about the establishment of robust nuclear industries, uninterrupted access to nuclear fuel cycles and fissile materials, and meeting major developmental, economic and climate change targets by placing nuclear at the centre of their energy security missions. In practice, this could also mean that the epicentre of the global nuclear industry could be shifting to the developing world, especially the bases in Asia, who could shape the future norms, best practices and structures for this domain.

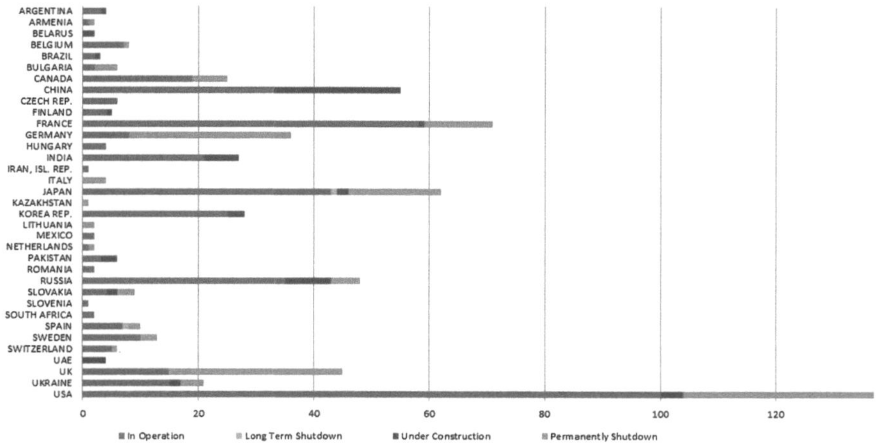

Fig. 3.1 Number of power reactors by country and status. *Source* IAEA (2016a)

According to estimates of the World Nuclear Association, around 45 countries are at various stages or plans for nuclear energy programmes, with 32 of them being in Asian region (WNA 2016e). This includes Albania, Italy, Serbia, Croatia, Portugal, Poland, Belarus, Norway, Estonia, Latvia, Turkey and Ireland (in Europe); Nigeria, Ghana, Senegal, Kenya, Uganda, Tanzania and Namibia (in Africa); and Cuba, Chile, Ecuador, Venezuela, Bolivia, Peru and Paraguay (in Americas). The gravity of the Asian shift is marked by the fact that nuclear projects in some subregions in Asia such as the Middle East and East Asia will be almost the same number or more than those planned in the above two regions. For example, 14 states in the Middle East (UAE and Saudi Arabia, Qatar, Kuwait, Yemen, Israel, Syria, Egypt, Tunisia, Libya, Algeria, Morocco, Jordan, and Sudan) have plans for setting up nuclear plants.

As the chart (Fig. 3.1) from IAEA shows, the United Arab Emirates (UAE) probably has the highest number of nuclear power project envisaged for a new entrant, with four units simultaneously been constructed (IAEA 2016c). In East Asia (including Australia and New Zealand in Oceania), almost ten nations (Indonesia, Philippines, Vietnam, Thailand, Laos, Cambodia, Malaysia, Singapore, Myanmar, North Korea) have nuclear energy plans afoot, with the North Korean interest in civilian use of nuclear energy stirring anxiety (thanks to its controversial military programme). The countries in Central Asia seeking nuclear energy include Azerbaijan, Kazakhstan,[3] Georgia and Mongolia, while in South Asia, Bangladesh

[3]While Kazakhstan plans to operationalise a Russian-built reactor by 2025, the Japanese government is seeking to extract the second site for a Japanese company. Industry observers expect an intense competition between Russian and Japanese companies for this project with both offering financing and technological support.

and Sri Lanka are seeking to follow the example of its larger neighbour, India, in placing reliance on nuclear power. Sri Lanka has already signed civil nuclear deal with India based on plans to start nuclear power generation by the year 2005 (WNN 2015). Bangladesh, for its part, plans two 1000 MW AES-92 nuclear plants at Rooppur, first to become operational by 2022. Another interesting aspect of the expectant renaissance is the countries such as Australia and Kazakhstan, which have among the world's largest uranium deposits and have eventually decided to pursue nuclear energy as an option and to utilise their own domestic resources. The same applies to Nigeria and Tanzania, which host large uranium deposits in Africa.

As the IAEA chart illustrates, the USA remains the country with the highest number of power reactors in operations and is placed at the top of the global supply chain. Other industrial majors such as France, Japan and Russia are close by being among the largest nuclear operators and their supplier behemoths maintaining global predominance. However, the actual story of the renaissance is about China, which easily outperforms these industrial powers in terms of the plants which are under rapid construction, and clearly indicating where the action is, provided Chinese companies also catch up in terms of reactor technology and figure prominently in the supplier industry. From the advanced world, it is only Russia's nuclear expansion plan that matches the Chinese strides, while the other prominent players in this burgeoning ecosystem are booming economies such as India and South Korea. Clearly, the inquisition is about how these Asian giants are going to buttress their nuclear industry towards a sustained growth trajectory that will create not just a strong Asian nuclear industrial domain but also revive fortunes of the global industry. While opportunities exist for these nations to fulfil these objectives, numerous hindrances exist in this path.

3.3.3 Systemic Factors as Stimulant

The optimism about a nuclear revival is also propped up by a handful of systemic stimulants. These include the advent of new generation of reactor technologies, improved access to and affordability of uranium as well as legal and risk management structures that intend to make the industry robust and accountable. Adding to this momentum is also a general corporate trend favouring consolidation within the global nuclear industry over the past decade.

3.3.3.1 Reactor Technologies

During the initial phase in the 1950s and 1960s, gas-/water-cooled and graphite-moderated designs formed the first generation of nuclear reactors (EuropeanCommission 2015). It was, however, the second generation of pressurised water (PWR) and boiling water reactor (BWR) that has dominated power generation since the 1970s and accounts for the bulk of reactors in operation

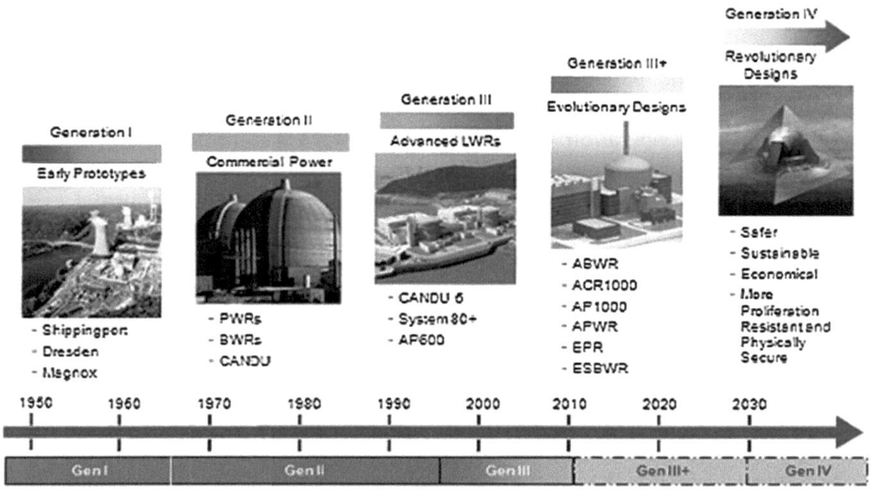

Fig. 3.2 Evolution of reactor technologies. *Source* European Commission. *Source* Chart available at: http://ec.europa.eu/research/energy/euratom/index_en.cfm?pg=fission&section=generation (accessed in September 2016)

worldwide. These include the PWR designs by Westinghouse, BWRs by General Electric, PHWRs by AECL (running on natural uranium), the Chinese CPR-1000 based on Framatome (Areva) design, the CNP PWRs of 600 and 1000 MW, the Korean Standard Nuclear Power Plant's (KSNP) OPR-1000s, (Goldberg and Rosner 2011), as well as the early variants of the Russian *Vodo-Vodyanoi Energetichesky* Reactors (VVER). Gen-II reactors were designed for a lifetime of 20–30 years with traditional safety features and passive engineering systems (Fig. 3.2).

A significant driver of the nuclear renewal is the advent of third generation (Gen-III) reactors, which comes with designs for extended lifetimes—up to 50–60 years—and enhanced safety and performance features. Available literature on Gen-III reactors describe them as having improved and standardised designs so as to expedite licensing, reduce cost and construction time, along with the advantage of their higher burn-up enabling efficient use of fuel and waste reduction (WNA 2016g). More importantly, they are designed to withstand conceivable safety challenges or any damage to the reactor that could result in core meltdown or radiological release. Advanced PWRs have capacities ranging from 600 to 1600 MW and the BWRs in the 1250–1550 range (FISA 2009). The prominent systems of this generation include Westinghouse's AP-600, General Electric (GE) ABWR, CANDU 6 and VVER-1000, among others (Goldberg and Rosner 2011).

While Gen-III designs have been in vogue since the mid-1990s, though only beginning to make notable market inroads, the factor that drives optimism in the industry is the rapid pace of technological upgrades as evident by the quick progress to Gen-III+ systems. Unlike the over 30-year gap between Generations II and

III, the galloping to the next technological level is visible in the Gen-III+ reactors which have integrated new benchmarks based on the basic Gen-III design. The significance for the expectant nuclear renewal is that the many ongoing and upcoming projects will be adopting Gen-III and III+ reactors, thus signifying a new era of nuclear plant operation. Gen-III+, in fact, comes with improvements in safety over Gen-III through passive safety features that do not require active controls or operator intervention (Marques 2011) to mitigate the impact of abnormal events. The systems on offer include the VVER-1200 (Rosatom), AP-1000/1400 (Westinghouse), Hualong 1 (CNNC) and Areva's EPR, considered the most advanced of this lot (Cognet 2010). However, it is vital to note that the prospects of a renaissance taking shape in this sector are heavily dependent on the success and wider acceptance of these systems and the technologies they claim to herald. Even the perceivably most advanced of these systems are yet to attain operational years or maturity amid aspersions being cast on various aspects of their designs, notably of EPR (Samuel 2015; Clercq and Mallet 2016) and AP-1000 (Sasi 2016b).

Nonetheless, despite the level of acceptability or success that these designs gain, the technological innovation will be an enduring process, going by the current pursuit of Generation-IV reactor technologies. Gen-IV hinges on six reactor technology concepts, namely gas-cooled and lead-cooled fast reactors, molten salt reactor, supercritical water-cooled reactor, sodium-cooled fast reactor and very high-temperature reactor (GIF 2014). According to WNA, the 13-nation strong Gen-IV International Forum intends to deploy some of these technologies, mainly fast neutron reactors, by 2020–2030, marking a new era of fast reactors and operating at higher temperatures (WNA 2016h). Besides this consortium, countries such as Russia (Patel 2016) and China (WNN 2014) are also working on waste-free fast neutron reactors, while those like India have long-running programmes on fast breeder reactors with the aim of harnessing plutonium and thorium resources. While assuming that Gen-III and III+ systems will drive nuclear revival and growth in the coming years, the future of the industry and its ability to remain as a sustainable source for the long term will invariably depend on the success attained by the Gen-IV development process.

3.3.3.2 A Liberal Fuel Access Regime

If reactor technologies are a key factor in shaping the expectant nuclear renaissance, a little lesser but significant role has also been played in this renewal process by the availability of or accessibility to nuclear fuel resources for the operating states, especially those newly setting up a nuclear energy infrastructure. One of the key aims of nuclear supplier states since the early days of creation of the non-proliferation regime has been to restrict the access of fissile materials, especially uranium and plutonium, from the reach of non-weapon states and certainly the states aspiring to develop nuclear weapons. Access to uranium has since the beginning been subject to IAEA safeguards with the movement of the material through the fuel cycle process being tracked at various stages.

On the other hand, uranium trading happens to be a vibrant mineral market with countries having huge deposits such as Canada, Australia, Kazakhstan and Brazil, among others, using their market dominance as a leverage to influence the rules in the supplier chain—involving both non-proliferation and pricing. While the global uranium market seems shaped by a balance of these two elements, the current state of the market indicates that both these elements may not always be complementary. Owing to various factors, global uranium prices were at a low for over a decade averaging $25–27 a ton (Aspa 2016b; Els 2016). Besides a general slump in the nuclear industry after the Fukushima incident also affecting the uranium market, oversupply is also a reason for the low prices. Production in mines worldwide has increased considerably over the years as major suppliers compete for dominance in the market share, thus adding to the oversupply (Aspa 2016a). The price slide has not been causal for more demand owing to cheaper alternatives like gas, which gave better options to power utilities. The market certainly looks to the states setting up new power plants, though most of them globally at still at construction stage, marking the interim phase before the expectant renaissance.

Beyond these factors, the market has also benefited from a non-proliferation-cum-disarmament dividend as global uranium inventories were supplemented by secondary sources such as recycled uranium and plutonium from used fuel, re-enriched depleted uranium, extracts from weapons-grade uranium and civil stockpiles, among others (WNA 2015b). While most market reports present a gloomy picture of uranium sector, the slump in prices might turn out to be promising news for nuclear operators and especially encourage aspirant states to invest further in setting up nuclear energy infrastructure. In that sense, the fall of uranium prices may prove to be a catalyst of global nuclear growth and enabling liberal access to fuel.

Despite the current lull in uranium prices, oversupply and liberal access to the yellow cake, supply may sooner or later diminish with depleting deposits and expectant restrictions on mining owing to environmental hazards. In this scenario, alternates like plutonium, if sufficiently secured against proliferation risks, could be seen as suitable means to reduce nuclear waste and harness the massive spent fuel stocks that have piled up worldwide. Further, fertile materials like thorium are yet to be substantially exploited and are available in more abundance than uranium deposits and could also be a key to future options as fuel that are proliferation-resistant and drastically reduce the amount of waste.

According to the World Nuclear Association, thorium could be introduced as fuel in seven types of reactors including the PHWRs, LWRs, BWRs, high-temperature gas-cooled reactors, accelerator-driven reactors, fast neutron reactors and molten salt reactors (WNA 2015c). Many countries including USA, Germany, Canada, France, Russia, China, Japan and India are working or have worked on reactor systems based on thorium. These include Germany's attempt to operate a thorium high-temperature reactor in the 1980s, the US efforts at Peach Bottom HTR in the 1960s, Fort St. Vrain HTR in 1970s and the light water breeder reactor at Shippingport in 1970s.

However, it is the Indian initiative to develop a fast breeder reactor (FBR) that stands out for its capability to breed thorium to produce U-233. Equally relevant is the fact that India is also known to have the world's largest thorium deposits. While other aspirants may maintain a latent interest in thorium, the centrality of thorium in India's three-stage programme creates hope that thorium will remain in contention as a long-term fuel option. Notwithstanding these efforts, a sense of pessimism prevails on the economic feasibility of the thorium route with not many nations convinced on betting it for the long term (Rahman 2011). The high cost and use of liquid sodium as coolant and the proliferation risks associated with large-scale presence of plutonium are key concerns that stymie the development of fast breeder programmes across the globe.

3.3.4 Risk Management Structures

Along with the evolution of the global nuclear industry, a set of institutions and instruments have come up over the years to manage various facets of the nuclear trade, especially to handle inherent risks and addressing issues of compensation and liability in the event of a nuclear incident that leads to radiation release and destruction. These instruments are crucial for the revival and further growth of the industry, considering the fact that public confidence has eroded by incidents like in Fukushima and the paranoia perpetuated by anti-nuclear groups around the world. Two significant developments in recent years that give fillip to the renaissance are the entry into force (IAEA 2015)[4] of the Convention on Supplementary Compensation for Nuclear Damage (CSC) and creation of nuclear insurance pools in emerging markets. The CSC builds upon two existing international conventions —the IAEA's Vienna Convention on Civil Liability for Nuclear Damage of 1963 (entered into force in 1977) and the Organisation of Economic Cooperation and Development's (OECD) Paris Convention on Third Party Liability in the Field of Nuclear Energy of 1960 (entered into force in 1968). Most European nations with nuclear energy infrastructure, as well as USA and Russia, are part of these conventions that form the existing regime for civil nuclear liability, besides the USA having the Price-Anderson Act as the world's first national nuclear liability legal framework.

Considering the global expansion of nuclear energy and trans-border implications of nuclear incidents, the CSC was drafted in 1997 as a comprehensive international liability framework with broad legal principles carried over from existing treaties. These include strict and exclusive liability of the nuclear operator, compensation without discrimination based on nationality, domicile or residence,

[4]The Convention stipulates at least 5 signatory states with a minimum of 400,000 units of installed nuclear capacity have to deposit their instrument of ratification with the IAEA for it can enter into force. Japan's ratification on 15 January 2015 triggered the Convention's entry into force three months later.

mandatory financial coverage of the operator's liability, exclusive jurisdiction (of courts of the state in which the nuclear accident occurs) and limitation of liability in amount and in time, among others (IAEA 2016b). One of the key objectives of CSC is to facilitate the availability of international supplementary funds through contribution of member states, which will place a larger amount at the disposal of the installation state to cater to compensation (IAEA 1998).

The CSC gives option to states to enact legislations having provision enabling the operator to exercise a right of recourse (RoR), wherein the cause of the incident/accident is attributed to the supplier. While most signatory states of these treaties have sought to indemnify the supplier, India's Civil Liability for Nuclear Damage Act of 2010 (BARC 2010) ran into controversy for its explicit references to supplier actions, which was resisted by both Indian and international suppliers (Kumar and Patil 2014; Balachandran 2015; Abraham and Mohan 2013). The Indian impasse was resolved through some governmental intervention that allowed flexibility of contracts between operator and suppliers to deal with RoR, besides establishing a national insurance pool to provide insurance to suppliers (MEA 2015). Even as the Indian law opened up new avenues of interpretation, many saw it as an emerging best practice that could set new standards for accountability and safety in this domain (Diksht 2013).

The establishment of a new international nuclear liability regime through the CSC is a timely addition to the global nuclear renewal efforts. Of all, the CSC provides the framework for signatory states to rationalise their national legal systems to a common set of principles while also providing for a set of standards and best practices for the aspirant states to adhere. Besides the availability of supplementary funds, the CSC provides for a benchmark of 300 million SDR which becomes the minimum requirement for a compensation framework for all installation states. While a CSC benchmark provides a minimum limit, rationalising global liability limits is difficult owing to the different size of nuclear industries in each nation. The CSC's supplementary funds, national liability funds or corpus created by governments to handle their part of liability (beyond operator liability) and national as well as international insurance pools together form the financial architecture that supports the global nuclear liability regime.

Nuclear pools are formed as underwriting forums to manage the insurance financial capacity of members' companies and provide for a large cash pool in order to address high-cost claims through their joint financial strength (Maciuli 2016). Owing to the high-intensity (though low-frequency) nature of nuclear incidents, the nuclear industry was not seen as a lucrative or safe sector by the insurance companies, which forced governments to encourage their national insurers to float such pools, and also tie up with international pools for better risk management and availability of funds. The American Nuclear Insurance Pool comprises over 60 investors, including foreign sources, and charges an average annual premium of $4 lakh for a reactor site (WNA 2016i). Over 26 insurance pools operate globally including the Nuclear Risk Insurers Ltd. (British pool), Assuratome (France), DKVG (Germany), the Japanese, Russian and Chinese pools, as well as the Nordic pool (Finland and Sweden combined) (NRI 2016).

Two cases of nuclear operating nations show how liability and insurance systems have emerged as key factors in shaping the contours of the global nuclear industry. For example, despite having the Atomic Energy Damage Compensation Law since 1961, the Japanese government had to substantially absorb the compensation burden as the operator liability was exempt in case of an incident arising from natural disasters. Japan passed a new legislation—Nuclear Damage Liability Facilitation Fund—in June 2011 to facilitate indirect assistance to the operator through the Nuclear Damage Compensation and Decommissioning Facilitation Corporation (NDF) by amortising bonds of up to 10 trillion yen (Endo 2016). Caught between the need to minimise taxpayer burden and ensure solvency of the operator, the Japanese government attempted various measures. These include converting Tepco (Fukushima operator) into a nationalised entity, devising a mutual-assistance scheme (involving 11 operators), promulgating a Corporate Rehabilitation Law and a dispute resolution framework, expanding government bond support to 9 trillion yen and allowing a scheme wherein burden is shared by operators, house-hold (power tariff), operators and government (public bonds).

The Japanese case is exemplar of how the sector could confront near-closure conditions if not backed by a risk management system that is financially and legally sound. The Indian case, on the other hand, is noteworthy for promulgating the civil liability law that provided for a substantial liability kitty without jeopardising operator and supplier interests. A struggle was visible on the insurance front where the state-run insurance companies, mandated to underwrite policies and form the national pool, struggled to create a system that could address the financial burden of the operators and suppliers, but also fulfil regulator guidelines. Further, the monopoly Indian operator, Nuclear Power Corporation of India Limited (NPCIL), was averse to international inspections which made it difficult for international insurers to access the Indian market (Official 2016). Eventually, the Indian pool came up in 2015, which facilitated the launching of an operator liability insurance policy (with limit of Rs. 1500 crore) and a Supplier Special Contingency Insurance (against right of recourse) Policy, among others. The Indian experience underlined the difficulties, emerging and new nuclear operating nations face, in establishing effective insurance and liability management structures, without which the creation of a robust nuclear industrial base could be difficult.

Other important instruments which have come up in recent years are (i) Convention on Early Notification of a Nuclear Accident, (ii) Convention on Assistance in the Case of a Nuclear Accident or Radiological Emergency, and (iii) Convention on Nuclear Safety. Industry groups have come together and have established World Association of Nuclear Operators (WANO) to provide peer-reviewed services. Peer reviews help members compare themselves against standards of excellence through an in-depth, objective review of their operations by an independent team from outside their organisation.

Another aspect that will impact the nuclear renewal process is the trend of consolidation witnessed in the global nuclear industry for more than a decade. Consolidation, involving mergers and acquisitions, buyouts and cross-ownerships, has been forced by the slump that the industry faced in the 1990s, including operational losses, depleting market share (in the surge of renewable sources) and technological development costs, among others. The last decade saw four major consolidation efforts, with two of them involving Western-Japanese mergers. Toshiba's majority ownership of Westinghouse and Hitachi semi-acquisition of General Electric's Nuclear Energy division (Hitachi 2007) followed an earlier European consolidation effort when France's Framatome and Cogema joined hands with Germany's Siemens to form Areva (GE 2015). Subsequently, Siemens exited this venture after Germany's nuclear phase-out plan, thus enabling French control over Areva and its subsequent tie-up with Mitsubishi of Japan along with acquiring entities in Sweden (Uddcomb) and USA (Duke). This consolidation process could be progressive as it enables strong integration of capital and technological bases, which could enable healthier growth trends and strengthening of the global market.

3.4 Asia Is Where the Hopes Lie

A key denominator to infer the probability of an impending renaissance is the evidence of revival or trend of substantial growth that could be witnessed globally or in a particular region. Though the developed world may have seen signs of revival, many nations remain hesitant to bet long term on nuclear and instead prefer to keep it as a 'saving grace' option in the event other green sources do not rise to expectations. So, is the nascent nuclear initiation plans in the developing world or swift expansion of nuclear projects among emerging economies testament to an impending renaissance? While these are all certain trends pointing to a renewed global interest in nuclear energy, the defining aspect of the renaissance narrative is the huge investments and reliance placed by the Asian powerhouses on nuclear.

In Asia, the nuclear industry is in an advanced evolutionary stage, marked by a quantum jump in number of reactors with a heavy concentration in three nuclear nations—China, India and South Korea (Japan already having a substantial number)—and a leap forward towards Gen-III and III+ technologies. That the epicentre of the industry is shifting to Asia is amply clear from these figures: of the 40 reactors under construction in Asia, 20 are in China, 6 in India and 3 each in South Korea and Japan; of the 89 projects planned, 40 are in China, 22 in India, 9 in Japan and 8 in South Korea. The world's leading nuclear suppliers have placed high hopes on these projects as their success could not just determine the viability of their new reactor technologies but also their own corporate future. However, each of the key players in Asia has their own inherent dynamics which might generate varied outcomes, one not necessarily favouring growth or renewal. When it comes to the key Asian players, the region could be divided into two groups: (a) Japan and South Korea as two small, but technologically advanced nations, seeking to make a windfall from the renaissance as leading technology providers and suppliers; (b) China and India, as the two most populous nations with high growth targets and energy consumption footprints, are destined to drive the growth as consumer states as also catalysts for future technology development.

3.4.1 Asia Techno-Leaders

Of the smaller two, South Korea already has 25 operating reactors (a number even a huge nation like India does not match) having an installed capacity of about 23,000 MW that is about 25% of its total installed electricity generating capacity. Seoul intends to raise the nuclear contribution to around 30% with a target of around 40,000 MW by 2030 (WNA 2016l). As a long-time nuclear energy producer and exporter, South Korea has been having a stable rate of growth and consistency in its nuclear business. Nonetheless, Seoul sees its role in impending renaissance as one of a technology leader and supplier. Like many advanced nuclear industrial bases, South Korea has developed its own Gen-III reactors—APR-1400 and OPR-1000 (based on its proprietary KSNP drawn from Westinghouse–Framatome technology). Besides using these systems for its own nuclear expansion, South Korea intends to deploy them to compete in the emerging markets of the developing world. Four APR-1400 form the core of Seoul's current contact to initiate UAE's nuclear programme, besides scanning interested markets in Southeast Asia. On the other hand, as a small nation with a huge penetration of nuclear power plants, South Korea also confronts the possibility of the public questioning its increasing nuclear imprint (Dalton and Cha 2016).

Japan, on the other hand, happens to be Asian nuclear giant on many attributes. (a) With over 40 operational reactors maintaining an installed capacity of close to 40,000 MW, Japan remains Asia's largest nuclear operating nation, likely to be supplanted by only China in the near future. (b) Japan had ended up as the world's leading nuclear supplier with its companies buying ownership in major US companies (Inagaki 2016) (Westinghouse–Toshiba and GE-Hitachi) and in Europe (Mitsubishi-EDF-Areva) (AFP 2016). As a technology leader in this domain, Japan sees a significant part of its nuclear future as a technology provider for emerging markets and aspirant nations. Tokyo is already pursuing nuclear cooperation agreements with India, Turkey and Kazakhstan and could be a major force in many other markets (Hanafusa 2015; Geropoulos 2016). Despite these attributes, the Japanese nuclear sector is at crossroads. The Fukushima incident led to closing down of all the nuclear plants even as the Japanese government under Shinzo Abe, a pronuclear politician, was struggling to restart these reactors amid intense public sentiments against nuclear energy (Kingston 2016).

Prior to the Fukushima incident, Japan's power sector was hugely dependent on nuclear energy, which was contributing nearly 30% of the total electricity generated. Japan had an ambitious plan to increase the nuclear contribution to over 40% by 2017 and up to 70% by 2030 (WNA 2016k). In fact, nuclear was emerging at the centre of Japan's energy security and carbon reduction strategies when Fukushima happened and derailed these plans. Following the nationwide shutdown of nuclear plants after the incident, Japan made an interim shift to coal-based power plants by setting up 45 new coal-fired plants to provide electricity. To bypass regulatory issues, size of a third of these plans was kept small, but it has raised questions about the cumulative effect on carbon emissions and how the country will meet its Kyoto

Protocol goals (Watanabe 2015). After importing over 80% of its energy sources in the form of coal, oil and natural gas, Japan has been forced to rethink its nuclear boycott and restore, even if partial, dependence on nuclear power (Bloomberg 2016).

Consequently, Japan's nuclear energy policy is at a critical juncture. Even while expanding its role and influence in the global nuclear industry, Japan is struck by a national dilemma of deciding what role nuclear will have in the nation's future. Following the 4th Basic Energy Plan formulated in December 2013, which values nuclear as a key base-load power source, Japan has initiated the process of restarting reactors one after another, with 3 starting and 2 others gaining approval till September 2016) (Stapczynski 2016). Even as observers predict around 20 reactors restarting by 2018, the continuance of local protests raises political costs for this policy, which though intends to decrease the dependence on nuclear energy over the years (WNN 2016a, b; ABCNet 2016). These incongruities raise questions not just on Japan's nuclear future, but also its own ability to contribute to the global nuclear renaissance.

3.4.2 China

3.4.2.1 Plans for Expansion

It may not be an overstatement to term China as the primary force that will drive the global nuclear energy renaissance. No other nation has the nuclear growth strategy that matches Beijing—besides 35 reactors in operation producing nearly 30 GW, 20 are currently under construction and at least 40 more in line (WNA 2016j). The 13th Five-Year Plan, from 2016 to 2020, envisages the building of around 6–8 new nuclear plants every year (only Russia comes anywhere close with a one large nuclear plant per-year policy) aiming to raise nuclear installed capacity to at least 58 GW by 2020–2021 and 150 GW by 2030 (Stone 2016). At that pace, some estimate that China will be having an installed nuclear capacity over 350 GW by 2050 with more than 400 reactors, far exceeding the USA (Conca 2015). However, this might sound overoptimism considering the numerous challenges the sector has been facing periodically, including issues of safety and waste management, which might not always favour a mass distribution of nuclear projects across the nation.

After spending considerable focus on the strategic weapon programme between the 1960 and 1980s, China was a comparatively late entrant into the civilian nuclear energy space, by conceiving nuclear energy projects in its 6th and 7th Five-Year Plans, and the first plant coming up only in 1991 (Zhou et al. 2011). Since then, Chinese public sector companies have being toiling through the learning curve to develop indigenous reactor technologies and the wherewithal for pursuing a closed fuel cycle after setting up the initial infrastructure with foreign assistance. From the 10th Plan onwards, ambitious projects intended at annual growth of the sector by 30% were formulated along with the twin focus on harnessing foreign collaboration and developing indigenous capabilities. It is, however, the 13th Plan of 2016 that signalled China's intention for a 'great leap forward' towards attaining global primacy in this sector—an objective that synchronises with its global aspirations. Owing to lesser political challenges as well as its impressive economic and technological surge, the Communist government have had the urge to raise the bar, constantly, by revising development targets. China's State Council in 2007 set a target of 40 GW in its 2005–2020 Nuclear Power Development Plan, which was subsequently raised to 60 GW, with expectations of expansion by at least 2 GW each year (CNR 2010).

The 13th Plan is a natural progression of this trend with rapid expansion targets including the deployment of Gen-III and III+ reactors. Key tenets of this Plan, according to WNA, include four AP1000 units at Sanmen and Haiyang, Hualong 1 reactors at Fuqing and Fangchenggang, demonstration CAP1400 reactor at Rongcheng (Shidaowan), Tianwan Phase III (units 5 and 6), a new coastal power plant, reaching target of 58 GW operational nuclear capacity by 2020 and building commercial reprocessing plants, among others (WNA 2016j). Going by developments of the past year, these ambitious plans seem very much on track with the sector witnessing numerous operational launches like the units in Yangjiang, Changjang and Fangchenggang—project sites where more units will be coming up in quick succession as most sites have capacity to hold multiple 1 GW reactors (WNN 2016a, b). According to some accounts, many of these projects in the range of 1000 MW have taken less than 5 years of construction time and cost only a third as that of Western nations (Conca 2015).

The quantum jump envisaged by the 13th Plan is supported by circumstantial requirements, namely the greater share earmarked for nuclear power in the total output and the quest for sustainable green energy sources. In fact, despite the prolonged pursuit of nuclear power infrastructure, its share remained minuscule in China's total electricity generation capacity. According to some analyses, nuclear power accounted for only 1% of installed capacity in 2009 (CNR 2010). By raising capacity to 60 GW, China intends to enhance the nuclear share to around 5% capacity by 2020. However, the nuclear expansion plan is not just about expanding share in the total output, but also about promoting nuclear as the centrepiece of China's green energy policies, which also includes investments in an energy mix consisting of renewable sources. As the world's most populous nation, and among the world's biggest polluters, at the centre of global climate change campaigns, China would prefer to place nuclear as the bulwark of its low-carbon policy and means to replace large-scale coal plants as a source for its electricity generation architecture.

3.4.2.2 Aiming for Technological Primacy

China's nuclear story is not just about more and more plants but also about its impressive technological spectrum. Despite heavy reliance on foreign technology, China, like in many other sectors, has developed competencies and attained, over the years, self-reliance in not just reactor design and construction, but also in other areas of the fuel cycle. With its record of reverse engineering, China proved its mettle in adapting Western technology, like Westinghouse AP-1000 becoming China's CAP-1000, as also developing indigenously designed systems like the Hualong 1 PWR. China's nuclear technology policy is two pronged: (a) develop Gen-III, III+ and IV reactors and (b) export its reactor as well as components and assemblies for other fuel cycle segments. As a nuclear weapon state under the Treaty on the Non-Proliferation of Nuclear Weapons (NPT), China had gained uninterrupted access to technologies and materials to build its nuclear industrial base since the early 1990s. China's subsequent membership in groups that control the nuclear trade, like the NSG, enabled its elevation as a lucrative market for new-generation reactors besides gaining liberal access to fuel supplies from key supplier markets (Parlez 2006; Bouw 2010).

The earliest Chinese power plants were based on Canadian, French and Russian designs, including Candu and AES-91, and subsequent indigenous Chinese improvisations such as CPR-1000 and CNP-300/600. Owing to substantial emphasis on nuclear growth, the shift to Gen-III and Gen-III+ reactors was rapid not just with international collaboration, but also efforts of Chinese companies to develop indigenous designs. Notable designs that made headway into the Chinese market include the EPR, AP-1000 and VVER-1000, among others. The competitive nature inherent in the functioning of China's public enterprises, often called as the 'nuclear troika'—the Chinese National Nuclear Corporation (CNNC), the China Guangdong Nuclear Power Corporation (CGNPC) and China Power Investment Corporation (CPIC)—has facilitated the progress achieved in localisation and indigenous development (Hinze and Zhou 2012). Western majors such as Westinghouse and Areva benefited from this competition by engaging the three Chinese companies. CNNC worked with Westinghouse and Framatome (Areva) to develop the CNP-1000, which was itself an improvement on the CNP-600/600, based on Areva's M310 design (WNA 2016j; Zhou et al. 2011). CNNC also deployed Westinghouse AP-1000 and went on to improvise itself Chinese version, CAP-1000 after signing a technology transfer agreement and augmenting capacity to 1400 MW. The CGNPC, on the other hand, hosted the Areva EPR as a Gen-III platform, which Areva claims to be of 4590 MW (thermal) and net electrical power of 1660 MW, besides developing the CPR-1000 as its indigenous version. The CPIC is the largest power investment organisation with investments in various sectors including nuclear projects covering AP-1000 and CPR-1000 systems as well as foreign projects.

While this colossal framework explains the current Chinese nuclear energy architecture, the other significant dimension of the Chinese nuclear story is its aspiration to be an influential player in the global nuclear market. Using its competencies in reactor design improvisation and development, China now offers the Hualong 1 in the international market and the main design for the projects it collaborates in foreign markets, like UK's Hinkley project. Hualong emerged out of a competition between CNNC and CGNPC to develop a 1000 MW for the export, with both designs being 'rationalised' and the ACP1000 and ACPR1000 'morphing' into the Hualong (WNA 2016j). While CNNC and CGN formed a joint venture to promote Hualong 1 in foreign markets, WNA reports that the export version is based on the CGN design, termed the HPR1000, as the CNNC's ACP1000 system has faced difficulties, and could be replaced by Hualong even in many domestic projects (Zhou et al. 2011; WNA 2016j).

While these dynamics of Gen-III system evolve, China is also investing heavily in the development of Gen-IV system including sodium-cooled fast neutron (Chinese experimental fast reactor), molten salt and gas-cooled (HTR), light water modular small reactors as well as thorium molten salt reactor for thorium exploitation. Going by the record of improvisation currently achieved, China is expected to make early headway in some of these technologies (Halper 2015). technologies. A harbinger of such optimism would be Chinese plans to develop and deploy small and modular reactors (SMRs) on various projects. Based on existing PWR designs, Chinese companies have competitively developed a series of SMRs, namely ACP100, CAP200 and ACPR100, among others. Even as various projects based on SMRs are already being pursued, the technology hit news after reports talked about Chinese plans to deploy SMRs in the disputed South China Sea region (Chen 2016). The Chinese nuclear story, in fact, is not about uninhibited growth alone, but also about occasional controversies. Like in many other countries, China too is facing waste management issues with constant public resistance to projects that involve reprocessing and nuclear waste disposal (Hornby 2016). There are also voices that question China's safety standards as well as the prudence of its massive nuclear expansion (Graham-Harrison 2015).

3.4.3 India

3.4.3.1 The Beginning

Though Indiadoes not match the growth witnessed in the Chinese nuclear domain, it figures as the other major player in the nuclear renaissance narrative by virtue of being the second most populous nation, and as one of the fastest growing economies, that has nuclear energy as the fulcrum of its green energy policy. There are unique attributes attached to India's nuclear programme which enables its high standing among the nations that could drive future growth of this sector. First, India was the only nation that made early breakthroughs in atomic technology but preferred to pursue a nuclear energy infrastructure instead of a weapon programme. India, as a key player in the establishment of the non-proliferation regime, played a decisive role in the creation of the IAEA (Fischer 1997) as well as the NPT negotiations. However, India refused to join the treaty and ended up undertaking a peaceful nuclear explosion (PNE) in 1974, following which international cooperation ceased and forced India to pursue a path of self-reliance in nuclear energy development. Very few nations in the developing world could match the infrastructural development that India pursued indigenously and forms the base for the future growth that could be undertaken along with renewed foreign participation. The history of the Indian programme is a narrative of how a state remains an outlier from the global normative structures and yet manages to traverse an independent path towards self-sufficiency.

Second, the architect of India's nuclear programme, Homi Bhabha, in his first note of April 1948 on the Organisation of Atomic Research in India, espoused the need to create an 'atomic pile' (IDSA 2016). Highlighting the scarce availability of uranium, Bhabha articulated the need for 'extensive and intense search for sources of uranium' as well as exploiting the abundant availability of thorium, while also harnessing international cooperation to kick-start the nuclear programme. This is significant considering that uranium has been a problematic factor that resonated throughout India's nuclear journey. On the other hand, the Indian programme has been in an evolutionary constant that seeks to transcend uranium dependence by leaping towards a thorium-based nuclear economy. The global future of thorium as a nuclear fuel (or fuel facilitator) is inevitably dependent on the success of India's thorium development efforts.

Third, India's three-stage programmes are widely considered as the 'most technically ambitious and innovative nuclear energy programme in the world (Rahman 2011)'. Bhabha announced plans for the three-stage programme at the second UN Conference on the Peaceful Uses of Atomic Energy in 1958 (Sreenivasan 2016). After using natural uranium for rectors in the first stage, the programme was supposed to deploy the plutonium from its spent fuel to run the fast breeder reactors (FBRs) in the second stage. Further, fertile thorium could be introduced as a blanket in the FBRs in order to produce uranium (U-233) that could eventually run the advanced boiled water reactor (ABWR) of the third stage. This implied a self-sustaining closed fuel cycle fostering a colossal nuclear infrastructure and enabling full exploitation of India abundant in thorium deposits. While announcing this programme, Bhabha had then proclaimed the goal of achieving 20,000 MW, though subsequently revised to 10,000 MW.

3.4.3.2 The Three-Stage Programme and Nuclear Renaissance

While talking about the India's potential to be a catalyst, alongside China, for a global nuclear energy renaissance, it is worth mentioning how India's nuclear programme is also about missing out on a home-grown renaissance. After the pursuit of Bhabha's vision began with Asia's first reactor, Apsara, in 1956, followed by Cirus (with US–Canada assistance), Bhabha's realisation that international support is needed to construct India's nuclear power infrastructure enabled agreements the Tarapur (USA) and Rajasthan (Canada) atomic power stations (IAEA 1971, 1974). However, following the cessation of international assistance after the 1974 PNE, India had to pursue indigenous development based on a pressurised heavy water reactor (PHWR) ecosystem. In fact, the third (1961–1966) and fourth (1969–1974) (DAE 1968) Five-Year Plan periods witnessed a plethora of R&D efforts including a research reactor (Zerlina), developing components for

PHWR reactors, setting up fuel fabrications complexes, reprocessing facilities, heavy water production, isotope production and applied electronics, among others. Two other significant forays were the uranium exploration project (in the Singhbhum Thrust Belt Area)[5] and research on fast breeders (construction of first FBR Purnima in 1972).

The turning point came in 1992 when the NSG, formed as a response to India's PNE, passed its 1992 guidelines stipulating nuclear supplies only for those countries with IAEA's full-scope safeguards (NSG 1992). While India struggled for more than a decade to gain fissile stocks and technology access, the 18 July 2005 India–US joint statement facilitated a 'nuclear deal' by which India was to be given an exclusive waiver from the NSG guidelines and allowed to access global nuclear commerce (DAE 2005). Accordingly, India got the waiver in September 2008, after it signed a new safeguard agreement with IAEA and presented a separation plan to place its civilian facilities under safeguards in a phased manner. A handful of major supplier nations, including France, Russia, USA, and subsequently Canada and Australia, among others, signs nuclear cooperation agreements with India, which enabled a steady flow of uranium into Indian reactors (PTI 2014).

The nuclear deal was promoted with the promise of initiating a sweeping expansion of the nuclear energy sector through new nuclear power plants constructed with foreign support, which could drastically increase nuclear energy production and enable a rapid progress towards 10,000 MW and beyond. At least six projects at following locations were announced during the nuclear deal, namely Jaitapur (Areva), MithiVirdi (initially Westinghouse, now status unknown), Kovvada (initially *GE*-Hitachi, now given to Westinghouse), Haripur (Rosatom), Kumharia (NPCIL), and Bargi/Chutkah and Pitti Sonapur (NPCIL-NTPC). The foreign-supplied reactors were of Gen-III, including EPR and AP-1000, with units of more than 1000 MW to build capacity of between of 2300 and 6000 MW in respective projects, which will have a cumulative contribution of over 25,000 MW and maybe more. On the other hand, there are 21 units currently in operation, endowing 5780 MW capacity (NPCIL 2016). Ongoing construction projects include four 700 MW units at 3 and 4 Kakrapar, and 7 and 8 of RAPS, unit 2 of Kudankulam Atomic Power Project, which will add another 3800 MW to the grid and taking crossing the 10,000 MW mark by end of this decade.

The logic behind the nuclear deal, which came amid intense domestic debate, is to use the assorted strength of foreign participation and indigenous PHWR projects, along with gains in the three-stage programme to not just meet the revised target, but the original Bhabha mission of 20,000 MW. The actual growth strategy though is beyond this goal, as evident from various declarations of achieving 450–600 MW by 2050, matching the Chinese plan (Sinha 2015). Such projections are based on the huge dividends expected from the three-stage programme. As a sequential process wherein the spent fuel of PHWRs is reprocessed into plutonium to run the

---

[5]Programme for Surveying, Prospecting and Development of Atomic Minerals during the IVth and Vth Plan periods (1969–1978) was released by Department of Atomic Energy.

FBRs, these reactors, besides producing power and converting the fertile thorium into U-233, also churn out more fuel that could return as fuel for the FBRs. While this creates the most self-sustainable fuel cycle that any operator could aspire, the other part of this revolutionary cycle starts when U-233 converted from thorium will run appropriately designed reactors and enable uninterrupted electricity generation, thanks to the abundant thorium deposits in the country.

Despite such promising plans, the programme remains stuck at the transition between the first and second stages. Efforts to progress to the second stage have overshot the initially envisaged timeline many times over. Though a 500 MW prototype fast breeder reactor (PFBR) at Kalpakkam was eventually expected to go critical by September 2015 (which though did not fructify) (Subramanian 2015), actual transition to a FBR ecosystem might take many more years. This contrasts the general perception that the second stage will formally begin with the commissioning of the PFBR. While the PFBR may undergo many processes, like preheating and core loading, before gaining approval from the Atomic Energy Regulatory Board (AERB), the progress towards higher power operations, criticality and eventual commissioning may depend on the functioning of the reactor during these processes. If launched in the next few months, the PFBR fuel cycle could run until 2021, after which construction of commercial FBRs (initially 1 and 2) may start by 2023–2024, with an expected completion time of 3–5 years. This implies that commercial power generation may start only around 2030. Further, the thorium blanket to generate fissile U233 for the next stage of reactors could be introduced only after the successful commercialisation of the FBRs (Kumar 2016).

Even if the second stage fructifies through the above given timeline, the key to the third stage rests on how sooner India's atomic establishment succeeds in developing Gen-IV designs including accelerator-driven system (ADS) (BARC 2007) and the advanced heavy water reactor (AHWR) (DAE 2008, 2009). Two different routes are known to be under development—the combinations involving Th-232 and Pu-239 and the other one involving Th-232 and U-233 (Jain 2008). Experimentations on the second route have been conducted at the KAMINI reactor in Kalpakkam, which is said to be the only operating reactor using U-233 as fuel (Jain 2008). The AHWR is designed and developed by Bhabha Atomic Research Centre (BARC) and has a unique combination of light water as coolant and heavy water as moderator. Going by development progress, the atomic establishment believes commercial construction of the first AHWR will begin soon and take many subsequent years for commissioning.

Questions were raised at the height of the nuclear deal debate on the prudence of opening up the Indian nuclear sector to foreign participation and whether this was necessitated by the frustrating delays in the three-stage programme. But for the constraints it faced in accessing uranium and Gen-III technologies, the Indian programme had made a significant headway in operating PHWRs, as also the wherewithal to run FBRs, which, many felt, could lead to a self-sustaining nuclear infrastructure (Kakodkar 2004). However, the government seemed clear in its plan of using foreign participation as a value-driven means to gallop towards unfulfilled as well as newer targets.

3.4.3.3 Impediments to India's Nuclear Renaissance

Even as this strategy unveils, there are other challenges hampering the prospects of a domestic nuclear renaissance in India. The foremost is the resistance spearheaded by anti-nuclear groups against nuclear energy in general and the expansion plans in particular. Following the Fukushima incident, a collective paranoia about the safety of nuclear plants and likely radiation fallouts have pervaded the populace, which resulted in widespread protests against existing projects like Kudankulam (DiaNuke 2012), as well as at new sites like Jaitapur (NDTV 2010). Encouraged by these movements, villagers in most planned sites have organised to block the planned projects, raising serious questions on the expansion plans (Shiva 2011). Adding to this momentum were political parties whose intervention had added to the uncertainty about these projects. Examples include the case of Jaitapur, where Shiv Sena is at loggerheads; Haripur, where the Chief Minister, Mamata Banerjee, had rejected the project, forcing the Russians to explore alternatives; and Kudankulam, where Chief Minister Jayalalitha's initial sympathy to the protesters had added traction to their movement.

Though the Fukushima event has been a catalyst for these movements, they also highlight the inefficacy of the nuclear establishment's socialisation efforts, especially when it comes to outreach and promotion of nuclear energy. That the establishment had remained aloof from the public with a veil of secrecy over its affairs had created difficulties in connecting to the communities, which were on the warpath (Kumar 2013). The Department of Atomic Energy, thereafter, launched campaigns to reach out and explain the safety features of India's nuclear power plants including its functional processes as well as issues such as nuclear waste management and disaster preparedness (IndianExpress 2011). How far these campaigns have made an impact could be known only when matters pertaining to land acquisition and reactor construction are going to be handled. Westinghouse's shifting to Kovvada after intense local opposition (Busvine and Jain 2016) and farmers clamouring for compensation at the Kovvada site are signs of the turmoil that is in store, with sufficient impetus to undermine India's nuclear expansion plans (Mahalakshmi 2016).

The other aspect is about demands for a much more transparent oversight over the nuclear industry through a nuclear regulatory authority. Various sections, as well as the IAEA, have called for an independent statutory regulator and also the delinking of the AERB, the current regulator from the nuclear establishment, through wider representation from different walks of the society, in order to make it more accountable (Sasi 2016a). The government introduced a Nuclear Safety Regulatory Authority Bill in 2011, which provided for replacing the AERB with a Nuclear Safety Regulatory Authority (NSRA). The Bill has now lapsed and the merits of opening up the sector are being intensely debated (Mohan and Kini 2016).

A related issue is about the lack of competitiveness in the Indian nuclear industry, unlike the Chinese case, with the role of nuclear operations being monopolised by state entities like NPCIL and BHAVINI. Most of the Indian structures—including regulatory, technological and risk management—are tailored to meet the operational principles of these state-run organisations. Nonetheless, many segments in the industry, including suppliers and insurers, hope for liberalisation of this sector in the near future, which could allow the entry of private operators, utilities and industrial consortiums that could provide a competitive edge and efficiency to this sector.

3.5 Conclusion

Renaissance is a perception or a sentiment that exists in the global nuclear industry —one of revival of an industry that has gone through various crises, and another of rejuvenation and renewal of a segment that have always seen sluggish and controlled growth, owing to the global politics and norms that influence the affairs of the atom. There have been numerous instances in the past when hopes of revival and renewal were bugled around but without fulfilling the factors that could have fructified a renaissance. Conditions have always been volatile enough to create mixed fortunes for the industry. While technology and low-carbon imperatives propelled the cause of nuclear industry, occasional incidents that send global alarm and persistent concerns over issues such as nuclear waste management and safety have consistently created insurmountable impediments for the industry.

The current wave of optimism and hopes of an impending nuclear renaissance is driven not just by the concerns of climate change, but also by the fact that many emerging economies have begun to consider nuclear as a sustainable source to power their economies. Similarly, finding newer markets is key to the future of global industry, including to promote its new-generation technologies. This is also important because the future of nuclear power itself is dependent on how sooner nations could shift to safer and modernised reactor systems along with harnessing various fuel sources while minimising waste and proliferation risks. Global mechanisms for risk management, including civil liability for nuclear damage and insurance, have also made a significant headway, thus enabling its evolution as a more credible and sustainable global commercial enterprise. Yet, the industry could not mature to a reliable business system which could run on vibrant buyer–seller dynamics or a comprehensive supply chain that could help further consolidation of the industry.

Despite many favourable factors and a facilitating environment, there are numerous spoilers that could delay or block the possibility of a complete renaissance. This chapter had indicated many spoilers though focussing a large part on the key players and the drivers that could propel the renewal or revival. Irrespective of the impact of these factors, a key determinant that could influence the industry's future is the level to which the general populace will adopt nuclear power. For, the key challenge to both revival and renewal is the lack of confidence among publics on the prudence of long-term dependence on nuclear energy. While government realistically push for nuclear as a sustainable energy source, the public industry disconnect remains a persistent issue that needs new strategies.

There are other policy and structural aspects as well that deserve greater attention. Foremost will be the question of whether nuclear power will be financially viable for smaller states for which only limited opportunities will be available to harness economies of scale. Will nuclear continue to remain an elitist preserve and be controlled by financially sound and industrially advanced nations, including the existing and the emerging ones? For the global nuclear industry that is desperately looking for newer markets and all avenues of growth, the need to be inclusive could drive some attention to the smaller states. Newer business models could be explored in order to further these causes, including areas such as joint ventures, public–private partnership and greater private ownership.

While these avenues are explored, the nuclear industrial future is also going to be influenced by how issues of safety and reliability are sufficiently addressed by newer reactors technologies including those that are currently being promoted under the Gen-III+ rubric as well as the futuristic ones that could come up as Gen-IV technologies. Similarly, it has to be seen how well new structures like CSC could find greater acceptance among the large section of countries operating nuclear power plants, and whether models adopted by countries like India will emerge as a new best practice. Lastly, irrespective of the revival in established bases and renewal promised in emerging segments, the prospects of nuclear renaissance are also going to be determined by how developed economies, especially though that are planning a phase-out, will approach nuclear power. Their return will mean an actual renewal!

References

ABCNet. (2016). *Japan reactor restarts, despite protests, boosting Tokyo's nuclear push* [Online]. Available at: http://www.abc.net.au/news/2016-08-12/japan-reactor-restarts-in-post-fukushima-nuclear-push/7729892. Accessed September 6, 2016.

Abraham, M., & Mohan, M. P. R. (2013). *Don't waver now on nuclear liability* [Online]. Available at: http://www.thehindu.com/opinion/lead/dont-waver-now-on-nuclear-liability/article5147177.ece. Accessed September 7, 2016.

AFP. (2016). Mitsubishi joins EDF in bid to save reactor builder Areva. *Mail Online*, July 28 [Online].

Aspa, J. (2016a). *10 top uranium mines* [Online]. Available at: http://www.u3o8.biz/s/MarketCommentary.asp?ReportID=765670&_Type=Market-Commentary&_Title=10-Top-Uranium-Mines. Accessed September 5, 2016.

Aspa, J. (2016b). *Uranium price forecast: Going up?* [Online]. Available at: http://investingnews.com/daily/resource-investing/energy-investing/uranium-investing/uranium-price-forecast/. Accessed September 1, 2016.

Balachandran, G. (2015). *Some issues in respect of Indian's nuclear liability law—I* [Online]. Available at: http://www.idsa.in/issuebrief/issuesinIndiansnuclearliabilitylaw_gbalachandran_100215. Accessed September 6, 2016.

BARC. (2007). *Accelerator driven system* [Online]. Available at: http://www.barc.gov.in/publications/eb/golden/reactor/toc/chapter8/8_1.pdf. Accessed September 30, 2016.

BARC. (2010). *Civil liability for nuclear damage act* [Online]. Available at: http://www.barc.gov.in/about/10.pdf. Accessed September 6, 2016.

Barrett, P. (2015). *Bloomberg Businessweek* [Online]. Available at: What Killed America's Climate-Saving Nuclear Renaissance? Accessed July 15, 2016.
BBC. (2016). Hinkley point: UK approves nuclear plant deal. *BBC*, September 15 http://www.bbc.com/news/business-37369786.
Bloomberg. (2016). *Japan's careful return to nuclear power* [Online]. Available at: https://www.bloomberg.com/view/articles/2016-03-10/japan-s-careful-return-to-nuclear-power. Accessed July 30, 2016.
Bouw, B. (2010). Camco signs long-term uranium supply deal with China. *The Globe and Mail*, Nov 24. http://www.theglobeandmail.com/globe-investor/cameco-signs-long-term-uranium-supply-deal-with-china/article4081015/.
Busvine, D., & Jain, R. (2016). Exclusive: Westinghouse to get new site for Indian nuclear plant—Officials. *Reuters*, May 31. http://www.reuters.com/article/us-india-westinghouse-idUSKCN0YM0L2.
CEW. (2015). *The history behind Germany's nuclear phase-out* [Online]. Available at: https://www.cleanenergywire.org/factsheets/history-behind-germanys-nuclear-phase-out. Accessed August 30, 2016.
Chen, S. (2016). *CNBC* [Online]. Available at: http://www.cnbc.com/2016/10/11/could-china-build-the-worlds-smallest-nuclear-power-plant-and-send-it-to-the-south-china-sea.html. Accessed September 3, 2016.
Clercq, G. D., & Mallet, B. (2016). *Reuters* [Online]. Available at: http://uk.reuters.com/article/edf-britain-idUKL8N15C22S. Accessed Aug 30, 2016.
CNR. (2010). A boost for nuclear power. *China Business Review*, May 1. http://www.chinabusinessreview.com/a-boost-for-nuclear-power/.
Cognet, G. (2010). *Generation 3 nuclear reactors* [Online]. Available at: http://www.iaea.org/inis/collection/NCLCollectionStore/_Public/44/078/44078364.pdf. Accessed July 5, 2016.
Conca, J. (2015). China shows how to build nuclear reactors fast and cheap. *Forbes*, October 22. http://www.forbes.com/sites/jamesconca/2015/10/22/china-shows-how-to-build-nuclear-reactors-fast-and-cheap/#9b2f5434d0b3.
DAE. (1968). *Allocation of plan provision for Vth plan for schemes* [Online]. Available at: http://idsa.in/npihp/documents/IDSA-IVPlan-18061968.pdf. Accessed September 1, 2016.
DAE. (2005). *Joint statement on India-USA civil nuclear co-operation*, July 18, 2005 [Online]. Available at: http://www.dae.nic.in/?q=node/61. Accessed September 1, 2016.
DAE. (2008). *Advanced heavy water reactor* [Online]. Available at: http://dae.nic.in/writereaddata/.pdf_37. Accessed September 30, 2006.
DAE. (2009). *Advanced heavy water system* [Online]. Available at: http://dae.nic.in/writereaddata/.pdf_31. Accessed September 30, 2016.
Dalton, T., & Cha, M. (2016). South Korea's nuclear energy future. *The Diplomat*, February 23 [online].
DiaNuke. (2012). *Latest updates from Koodankulam* [Online]. Available at: http://www.dianuke.org/latest-updates-from-koodankulam/. Accessed September 20, 2016.
Diksht, S. (2013). *Japan may amend its nuclear damage compensation Act* [Online]. Available at: http://www.thehindu.com/sci-tech/energy-and-environment/japan-may-amend-its-nuclear-damage-compensation-act/article4476106.ece. Accessed July 7, 2016.
Els, F. (2016). *Uranium market is getting crushed* [Online]. Available at: http://www.mining.com/uranium-market-getting-crushed/. Accessed September 5, 2016.
Endo, N. (2016). *The nuclear damage compensation and decommissioning facilitation corporation (NDF)* [Online]. Available at: http://www.ndf.go.jp/soshiki/pamph_e.pdf. Accessed September 3, 2016.
EnergyPost. (2016). *The looming Nordic energy crisis* [Online]. Available at: http://energypost.eu/looming-nordic-energy-crisis/. Accessed August 1, 2016.
EnergyTransition. (2015). *The global energy Wende* [Online]. Available at: https://energytransition.org/2015/08/gersman-nuclear-phase-out-is-economically-sensible/. Accessed July 1, 2016.

EuropeanCommission. (2015). *European Commission* [Online]. Available at: http://ec.europa.eu/research/energy/euratom/index_en.cfm?pg=fission&section=generation. Accessed August 30, 2016.

FISA. (2009). *GenIII/III+ nuclear reactors*. TVO: Prague.

Fischer, D. (1997). *History of the international atomic energy agency*. Vienna: IAEA.

GE. (2015). *GE completes acquisition of Alstom power and grid businesses* [Online]. Available at: http://www.genewsroom.com/press-releases/ge-completes-acquisition-alstom-power-and-grid-businesses-282159. Accessed September 30, 2016.

Geropoulos, K. (2016). Energy hopes, US drive renewed Turkey, Israel relations. *New Europe*, July 1 [Online].

GIF. (2014). *Technology roadmap update for generation IV nuclear energy systems*. Paris: GIF.

Goldberg, S. M., & Rosner, R. (2011). *Nuclear reactors: Generation to generation*. Cambridge: American Academy of Arts and Sciences.

Graham-Harrison, E. (2015). China warned over 'insane' plans for new nuclear power plants. *Guardian*, September 2016. https://www.theguardian.com/world/2015/may/25/china-nuclear-power-plants-expansion-he-zuoxiu.

Guardian. (2016). *Flamanville: France's beleaguered forerunner to Hinkley Point C* [Online]. Available at: https://www.theguardian.com/environment/2016/jul/27/flamanville-france-edf-nuclear-reactor-hinkley-point-c. Accessed August 1, 2016.

Halper, M. (2015). *The U.S. is helping China build a novel, superior nuclear reactor* [Online]. Available at: http://fortune.com/2015/02/02/doe-china-molten-salt-nuclear-reactor/. Accessed September 3, 2016.

Hanafusa, R. (2015). Turkey to build 4 reactors by 2028 with Japan's help. *Nikkei Asian Review*, March 21 [Online].

Hinze, J., & Zhou, Y. (2012). *NEI magazine China supplement (Belfer Centre)* [Online]. Available at: http://belfercenter.hks.harvard.edu/files/chinas-commercial-reactors-proof.pdf. Accessed September 8, 2016.

Hitachi. (2007). *Hitachi* [Online]. Available at: http://www.hitachi.com/New/cnews/f_070516b.pdf. Accessed September 30, 2016.

Hornby, L. (2016). China's protests force rethink on nuclear waste site. *Financial Times*, August 10. https://www.ft.com/content/003ecb60-5ec7-11e6-bb77-a121aa8abd95.

IAEA. (1971). *The text of the safeguards agreement relating to the bilateral agreement between India and the United States of America* [Online]. Available at: https://www.iaea.org/sites/default/files/publications/documents/infcircs/1971/infcirc154.pdf. Accessed September 10, 2016.

IAEA. (1974). *The text of a safeguards agreement between the agency, Canada and India* [Online]. Available at: https://www.iaea.org/publications/documents/infcircs/text-safeguards-agreement-between-agency-canada-and-india. Accessed September 10, 2016.

IAEA. (1998). *Convention on supplementary compensation for nuclear damage* [Online]. Available at: https://www.iaea.org/publications/documents/infcircs/convention-supplementary-compensation-nuclear-damage. Accessed September 5, 2016.

IAEA. (2015). *Convention on supplementary compensation enters into force* [Online]. Available at: https://www.iaea.org/newscenter/news/convention-supplementary-compensation-nuclear-damage-enters-force. Accessed September 6, 2016.

IAEA. (2016a). *International atomic energy agency* [Online]. Available at: https://www.iaea.org/PRIS/CountryStatistics/CountryStatisticsLandingPage.aspx. Accessed August 30, 2016.

IAEA. (2016b). *Vienna convention on civil liability for nuclear damage* [Online]. Available at: https://www.iaea.org/publications/documents/conventions/vienna-convention-on-civil-liability-for-nuclear-damage. Accessed September 5, 2016.

IAEA. (2016c). *Number of power reactor by country and status* [Online]. Available at: https://www.iaea.org/PRIS/CountryStatistics/CountryStatisticsLandingPage.aspx. Accessed July 5, 2016.

IDSA. (2016). *Nuclear history* [Online]. Available at: http://www.idsa.in/npihp/documents/IDSA-HBP-26041948.pdf. Accessed August 20, 2016.

Inagaki, K. (2016). Westinghouse deal is a welcome bright spot for Toshiba. *Financial Times*, June [Online].

IndianExpress. (2011). *Indian Express* [Online]. Available at: http://indianexpress.com/article/news-archive/latest-news/kalam-suggests-10point-action-plan-on-koodankulam-project/. Accessed September 30, 2016.

Jain, S. (2008). *Nuclear power—An alternative* [Online]. Available at: http://www.npcil.nic.in/pdf/nuclear%20power-%20an%20alternative.pdf. Accessed September 30, 2012.

Kakodkar, A. (2004). *Fast-breeder reactors more important for India* [Interview]. November 24, 2004.

Kingston, J. (2016). *China and East Asia: September 2016*. Philadelphia: Current History.

Kumar, A. V. (2013). India's nuclear energy renaissance: Stuck in the middle? *Journal of Risk Research, 17*(1), 43–60.

Kumar, A. V. (2016). *India-US nuclear deal: Only half realised after a decade* [Online]. Available at: http://www.idsa.in/idsacomments/IndiaUSNuclearDeal_avkumar_210715. Accessed September 30, 2016.

Kumar, A. V., & Patil, K. (2014). *Resolving India's nuclear liability impasse* [Online]. Available at: http://www.idsa.in/issuebrief/ResolvingIndiasNuclearLiabilityImpasse_kumarpatil_061214. Accessed September 6, 2016.

Maciuli, I. (2016). *Nuclear Niche: Demystifying nuclear insurance* [Online]. Available at: https://international.ig.cpcusociety.org/news/nuclear-niche-demystifying-nuclear-insurance. Accessed September 2016.

Mahalakshmi, B. (2016). Despite defiant farmers, Andhra govt hopes to complete Kovvada nuclear project land survey soon. *The financial express*, June 11. http://www.financialexpress.com/economy/despite-defiant-farmers-andhra-govt-hopes-to-complete-kovvada-nuclear-project-land-survey-soon/281079/.

Marques, J. G. (2011). Review of generation-III/III+ fission reactors. In S. Krivit, J. Lehr, & T. Kingery (Eds.), *Nuclear energy encyclopedia: Science, technology, and applications* (pp. 231–254). New York: Wiley.

MEA. (2015). *Frequently asked questions and answers on civil liability for nuclear damage act 2010 and related issues* [Online]. Available at: http://www.mea.gov.in/press-releases.htm?dtl/24766/Frequently._Asked_Questions_and_Answers_on_Civil_Liability_for_Nuclear_Damage_Act_2010_and_related_issues. Accessed July 30, 2016.

Meyer, G., Hume, N., & Sheppard, D. (2016). *Financial Times* [Online]. Available at: https://www.ft.com/content/3bc0116c-e681-11e5-a09b-1f8b0d268c39. Accessed August 1, 2016.

Milne, R. (2016a). *Financial Times* [Online]. Available at: https://www.ft.com/content/b44e3214-2f13-11e6-bf8d-26294ad519fc. Accessed August 2, 2016.

Milne, R. (2016b). *Financial Times* [Online]. Available at: https://www.ft.com/content/909e2980-2740-11e6-8ba3-cdd781d02d89. Accessed August 1, 2016.

Mohan, M. R., & Kini, E. R. (2016). India's nuclear regulators have been audited. *Business Line*, January 3. http://www.thehindubusinessline.com/opinion/indias-nuclear-regulators-have-been-audited/article8061473.ece.

NDTV. (2010). *Protests over Jaitapur* [Online]. Available at: http://www.ndtv.com/topic/protests-over-jaitapur. Accessed September 30, 2016.

NEI. (2016). *Nuclear energy institute* [Online]. Available at: http://www.nei.org/News-Media/News/News-Archives/nrgenergyfilesfirstfulllicenseapplicationfornewrea. Accessed July 1, 2016.

NewEurope. (2016). *New units said safer, less radioactive waste* [Online]. Available at: https://www.neweurope.eu/article/world-retreats-nuclear-power-russia-pushes-faster-reactors/. Accessed July 3, 2016.

NPCIL. (2016). *Nuclear power generation 2010-16* [Online]. Available at: http://www.npcil.nic.in/main/AllProjectOperationDisplay.aspx. Accessed September 3, 2017.

NRI. (2016). *Nuclear risk insurers* [Online]. Available at: http://www.nuclear-risk.com/. Accessed July 30, 2016.

NSG. (1992). *Press statement of nuclear suppliers meeting* [Online]. Available at: http://www.nuclearsuppliersgroup.org/images/Files/Documents-page/Public_Statements/1992-Press.pdf. Accessed September 2, 2016.

Official, I. (2016). *Nuclear* [Interview]. July 30, 2016.
Parlez, J. (2006). China signs deal for Australian uranium. *The New York Times*, April 2. http://www.nytimes.com/2006/04/02/world/asia/china-signs-deal-for-australian-uranium.html.
Patel, S. (2016). *Power Magazine* [Online]. Available at: http://www.powermag.com/russian-fast-reactor-connected-grid/. Accessed August 30, 2016.
PTI. (2014). *Over 4,000 tonnes uranium imported for n-reactors since 2008* [Online]. Available at: http://www.business-standard.com/article/pti-stories/over-4000-mt-uranium-imported-for-nuclear-reactors-since-2008-114121100742_1.html. Accessed September 11, 2016.
Rahman, M. (2011). *Homi Baba's vision turned India into a nuclear R&D leader* [Online]. Available at: http://www.guardian.co.uk/environment/2011/nov/01/homi-bhabha-india-thorium-nuclear?intcmp=239. Accessed September 5, 2016.
Riley, C., & Mullen, J. (2016). China will own part of new U.K. nuclear power plant. *CNN Money*, September 15. http://money.cnn.com/2016/09/15/news/china-uk-nuclear-power-plant-hinkley-point/.
Samuel, H. (2015). *The Telegraph* [Online]. Available at: http://www.telegraph.co.uk/news/earth/energy/nuclearpower/11727000/Areva-aware-as-early-as-2006-of-serious-fault-in-nuclear-reactor-destined-for-UK.html. Accessed August 30, 2016.
Sasi, A. (2016a). *Indian Express* [Online]. Available at: http://indianexpress.com/article/india/india-others/new-atomic-regulator-nuking-the-autonomy-red-flag/. Accessed September 30, 2016.
Sasi, A. (2016b). *Indian Express* [Online]. Available at: http://indianexpress.com/article/business/business-others/westinghouse-ap1000-reactors-patchy-record-cost-concerns-loom-large-2867947/. Accessed August 30, 2016.
Shiva, V. (2011). *Nuclear Insanity* [Online]. Available at: http://www.globalresearch.ca/nuclear-insanity/24963?print=1. Accessed September 15, 2016.
Sinha, A. (2015). Anil Kakodkar: Must target 63,000 MW nuclear-power by 2032. *Indian express*, February 4. http://indianexpress.com/article/india/india-others/anil-kakodkar-must-target-63000mw-nuclear-power-by-2032/.
Sreenivasan, M. (2016). *Former chairman atomic energy commission* [Online]. Available at: http://dae.nic.in/?q=node/220. Accessed September 2, 2016.
Stapczynski, S. (2016). *Japan reactor restart signals latest step in nuclear rebirth* [Online]. Available at: https://www.bloomberg.com/news/articles/2016-08-11/japan-reactor-restart-signals-latest-step-in-nuclear-rebirth. Accessed September 10, 2016.
Stone, J. (2016). China to build 40 nuclear power plants over the next five years. *The independent,* January 4. http://www.independent.co.uk/news/uk/politics/china-to-build-40-nuclear-power-plants-over-the-next-five-years-a6795401.html.
Subramanian, T. (2015). Kalpakkam breeder reactor to go on stream. *The Hindu*, July 8. http://www.thehindu.com/sci-tech/science/kalpakkam-prototype-fast-breeder-reactor-to-go-on-stream/article7396541.ece.
Tindale, S. (2011). *Thorium: How to save Europe's nuclear revival*. London: Centre for European Reform.
Watanabe, C. (2015). *Japan coal revival ducks scrutiny with small plants, critics say* [Online]. Available at: http://www.bloomberg.com/news/articles/2015-05-27/size-matters-as-japan-coal-revival-ducks-scrutiny-due-big-plants. Accessed September 10, 2016.
WHO. (2016). *World Health Organisation* [Online]. Available at: http://www.searo.who.int/india/topics/sustainable-development-goals/SDG_health/en/. Accessed May 2, 2016.
WNA. (2015a). *The Nuclear Renaissance* [Online]. Available at: http://www.world-nuclear.org/information-library/current-and-future-generation/the-nuclear-renaissance.aspx. Accessed July 20, 2016.
WNA. (2015b). *Uranium Markets* [Online]. Available at: http://www.world-nuclear.org/information-library/nuclear-fuel-cycle/uranium-resources/uranium-markets.aspx. Accessed September 5, 2016.
WNA. (2015c). *Thorium* [Online]. Available at: http://www.world-nuclear.org/information-library/current-and-future-generation/thorium.aspx. Accessed September 5, 2016.

WNA. (2016a). *US Operating Nuclear Reactors* [Online]. Available at: http://world-nuclear.org/information-library/country-profiles/countries-t-z/appendices/nuclear-power-in-the-usa-appendix-1-us-operating-n.aspx. Accessed August 1, 2016.
WNA. (2016b). *Nuclear Power in the USA* [Online]. Available at: http://world-nuclear.org/information-library/country-profiles/countries-t-z/usa-nuclear-power.aspx. Accessed August 12, 2016.
WNA. (2016c). *Nuclear Power in France* [Online]. Available at: http://www.world-nuclear.org/information-library/country-profiles/countries-a-f/france.aspx. Accessed August 1, 2016.
WNA. (2016d). *Nuclear Power in Finland* [Online]. Available at: http://www.world-nuclear.org/information-library/country-profiles/countries-a-f/finland.aspx. Accessed August 1, 2016.
WNA. (2016e). *Emerging Nuclear Energy Countries* [Online]. Available at: http://www.world-nuclear.org/information-library/country-profiles/others/emerging-nuclear-energy-countries.aspx. Accessed September 30, 2016.
WNA. (2016f). *Nuclear Power in Russia* [Online]. Available at: http://www.world-nuclear.org/information-library/country-profiles/countries-o-s/russia-nuclear-power.aspx. Accessed August 1, 2016.
WNA. (2016g). *Advanced Nuclear Power Reactors* [Online]. Available at: http://www.world-nuclear.org/information-library/nuclear-fuel-cycle/nuclear-power-reactors/advanced-nuclear-power-reactors.aspx. Accessed September 30, 2016.
WNA. (2016h). *Generation IV Nuclear Reactors* [Online]. Available at: http://www.world-nuclear.org/information-library/nuclear-fuel-cycle/nuclear-power-reactors/generation-iv-nuclear-reactors.aspx. Accessed August 30, 2016.
WNA. (2016i). *Liability for Nuclear Damage* [Online]. Available at: http://www.world-nuclear.org/information-library/safety-and-security/safety-of-plants/liability-for-nuclear-damage.aspx. Accessed September 10, 2016.
WNA. (2016j). *Nuclear Power in China* [Online]. Available at: http://www.world-nuclear.org/information-library/country-profiles/countries-a-f/china-nuclear-power.aspx. Accessed September 10, 2016.
WNA. (2016k). *Nuclear Power in Japan* [Online]. Available at: http://www.world-nuclear.org/information-library/country-profiles/countries-g-n/japan-nuclear-power.aspx. Accessed September 7, 2016.
WNA. (2016l). *Nuclear Power in South Korea* [Online]. Available at: http://www.world-nuclear.org/information-library/country-profiles/countries-o-s/south-korea.aspx. Accessed September 30, 2016.
WNN. (2014). *World Nuclear News* [Online]. Available at: http://www.world-nuclear-news.org/NN-Chinese-fast-reactor-completes-full-power-test-run-1912144.html. Accessed August 30, 2016.
WNN. (2015). *India, Sri Lanka agree to nuclear cooperation* [Online]. Available at: http://www.world-nuclear-news.org/NP-India-Sri-Lanka-agree-to-nuclear-cooperation-1602154.html. Accessed July 5, 2016.
WNN. (2016a). *Japan institute sees 19 reactors restarts by March 2018* [Online]. Available at: http://www.world-nuclear-news.org/NP-Japanese-institute-sees-19-reactor-restarts-by-March-2018-2807164.html. Accessed July 2, 2016.
WNN. (2016b). *Three Chinese units approach operation* [Online]. Available at: http://www.world-nuclear-news.org/NN-Three-Chinese-units-approach-operation-0407164.html. Accessed September 10, 2016.
Zhou, Y., Rengifo, C., Chen, P., & Hinze, J. (2011). Is China ready for its nuclear expansion? *Energy Policy, 39*(29), 771–781.

Author Biography

Mr. A. Vinod Kumar is Associate Fellow at the Institute for Defence Studies and Analyses (IDSA), and a Visiting Faculty at the Institute of Foreign Policy Studies (IFPS), University of Calcutta, Kolkata. His research interests include nuclear policy issues (including non-proliferation, nuclear energy and deterrence), missile defence and India's relations with the great powers. Kumar's first book titled *India and the Nuclear Non-Proliferation Regime—The Perennial Outlier* was published by the Cambridge University Press in April 2014. He has written extensively on these issues, as well as national and international politics, in acclaimed international publications. Prior to joining IDSA, Kumar was a journalist, with stints as a private broadcaster as well as a first generation online journalist. Kumar was earlier a Fellow at the Indian Pugwash Society. He is now spearheading an archival mining effort to trace India's nuclear and foreign policy history.

Chapter 4
Nuclear Power and Climate Change Mitigation: Search for Low-Carbon Energy Mix in Asia

Nandakumar Janardhanan

4.1 Introduction

Asia's demand for energy is expected to grow significantly in the decades to come to fuel its growth. It is projected that the developing Asia alone could account for 44% of global GDP by 2035 with 6% annual economic growth (ADB 2013b). This undeniably indicates an enormous growth in the energy demand by the region during the coming decades. This period will also witness majority of the energy demand from India and China, where fossil fuel will remain to be the largest consumed energy sources. It is estimated that coal consumption would increase by 81%, oil consumption will be doubled, and natural gas consumption will be tripled by 2035 (ADB 2013b). The surge in energy consumption will also lead to increasing energy-related emissions which is estimated to reach roughly over 21 billion tons though the global emission needs to be kept under 22 billion tons to ensure that GHG emissions are in balance with the planet's health (ADB 2013b). Without major changes in the energy policy of Asian countries, the existing fuel mix will not change.

One of the most interesting aspects of the energy scenario in developing Asia is that despite the increase in consumption, significant share of population in these regions will continue to remain without having access to electricity. The Asian Development Bank estimates that except China, all other countries in the region will face electricity supply deficit by 2030. Moreover, biomass will remain to be a major source of energy for cooking and energy-related emissions will substantially increase which in turn can adversely affect human as well as environmental health among Asian countries.

Given the above energy background, balancing energy demand growth and environmental objective is a critical policy challenge to all the countries in

N. Janardhanan (✉)
Energy Studies Programme, Jawaharlal Nehru University, New Delhi, India
e-mail: nanduj123@gmail.com

N. Janardhanan et al. (eds.), *Resurgence of Nuclear Power*,
https://doi.org/10.1007/978-981-10-5029-9_4

developing Asia. While some of the major economies in the regions are already considered as the top emitters, many more countries are emerging as critical contributors of GHG emissions.

Countries such as China and India have made significant progress in alternative and renewable energy development. New energy technologies have also been explored and are being tested for their feasibility at the commercial level. Several studies focusing on the region's energy scenario have highlighted that there are significant opportunities for non-fossil fuels, especially nuclear. This chapter will be focusing on climate mitigation and energy choices of Asian economies and would highlight the opportunities for nuclear power. The details included in this chapter are limited to developing Asian countries.

4.2 Asia's Energy Security

Two sets of narratives on energy security debates in Asia will be discussed in this section. The first strand of narrative highlights the argument that energy security is purely a rational policy decision by individual countries that focuses ultimately on the national interest and for fuelling the domestic economic activities. This narrative has been the prevailing point in the energy security debates of many of the Anglo-Saxon economies which saw energy as a commodity crucial to national security. The Anglo-Saxon societies have considered oil as one of the primary energy sources for meeting its national security objectives. The Anglo-Saxon search for oil has also been linked closely to the battles that were fought until recent decades. Energy always played a critical role in the 'national security' front as manufacturing of weapons and military movements demand huge energy inputs. Experts observe that many 'modern wars are waged with weaponry whose construction requires some of the most energy-intensive materials and whose deployment relies on incessant flows of secondary fossil fuels (e.g. gasoline, kerosene) and electricity to energise the machines that carry them and to equip and provision the troops who operate them' (Smil 2004). As mentioned earlier, the ability of petroleum fuel—oil and gas—to undergo combustion in controlled manner and produce heat or other usable forms of energy has made the same preferred fuel sources in many countries.

The second strand of debate highlights that energy policy of a country must be in tune with that of the regional as well as global energy–environment goals and must keep evolving. This debate gained attention among policy making across the world in the post-1972 period after the UN Conference of Human Environment held in Stockholm. However, the debate on energy–environment linkages was overshadowed by the 1973 Arab oil embargo and the subsequent oil shortages that happened during Iranian Revolution. Though energy diversification gained attention in the immediate following months of Arab oil embargo in major energy consuming countries, the fall in oil price in the subsequent years made the diversification strategies too costly to continue.[1] Many of the energy investments that were directed towards renewable energy development proved not to be economically viable as oil price plunged. To put in perspective, oil price which was $3 per barrel prior to 1973 skyrocketed to $12 per barrel during the Arab oil embargo; however, in the post-embargo period, price fell back close to the original level. This fluctuation was a bigger challenge to non-conventional energy sector. Moreover, global energy market began to be controlled again by the security perception, especially due to the fact that majority of the world resources of oil and gas are located in the Persian Gulf region which has been witnessing geopolitical turbulence.

A change in perception began in the post-Gulf War period with the United Nations Framework Convention on Climate Change (UNFCCC) held in 1992 which aimed at discussing the adverse effects of greenhouse gases on the global warming. Anthropogenic emissions have been identified as one of the major reasons behind the global warming happening in the modern-day period, and commitment to reduce the same by countries was necessary. The conclusion of Kyoto Protocol in 1997 (UNFCCC 1997) was another critical step that changed the course of narrative of conventional energy security perception, where fossil fuels alone were viewed as the key source of energy supply for running the global economic engine. Further, last decade witnessed significant changes in the debate of energy security across the world. The USA led invasion in Iraq, and the subsequent geopolitical challenges in Persian Gulf region evinced various implications for the global energy market. Building 'Strategic Petroleum Reserves (SPR)' to enhance stock pile capacity, conducting energy search outside Middle East and promoting investment in non-conventional energy were major policy measures adopted by many import dependent countries. All these have contributed to rise of oil price, from $28 per barrel in 2003 to $148 per barrel by 2008. However, the subsequent economic downturn led to the falling demand and eventually led to stabilising price at around $80 per barrel.

It is important to note that all these developments have contributed to better investment in non-fossil fuel sector. Here, another major challenge emerged. The intermittency in the supply of renewable energy and the demand for huge investment required for developing these energy facilities as well as storage concerns put a cap on the attractiveness of the same. In today's world, countries have limited choices of energy sources that are free from political, economic, technological, or environmental limitations. However, the demand for faster economic growth, need for poverty alleviation, and ensuring cleaner energy access highlight that transitioning from a conventional fossil fuel-dominated energy mix to a newer mix with cleaner energy sources is important to balance economic goals with environmental objectives.

---

[1]The diversification strategies mainly consisted of promotion of renewable or other non-conventional sources.

4.3 Emission Reduction Targets in Asian Region

The potential growth in the energy consumption in Asia highlights the fact that the continuing dependence on fossil fuel could bring serious challenges to the environment. Unlike the Western world where advanced technology has been playing a key role in addressing the energy-related emissions, the developing economies in Asia face the dual challenges due to over dependence on fossil fuels. First, the fossil consumption directly contributes to GHG emissions and thereby affects environment. Second, being agrarian economies, growing GHG emissions also pose significant challenges to livelihood and food production in developing Asia (Fig. 4.1).

Among the developing Asia, China and India have been the two largest GHG emitters consistently for several years. Though alternative energy sources are increasingly growing in the domestic energy mix, the share of the same would not be able to replace the conventional fuel sources. The consumption of coal, oil, and natural gas will increase significantly in the region in the years to come if the current consumption pattern continues. It is estimated that commercially traded primary energy consumption of nine countries in the developing Asia including 'China, India, Indonesia, Thailand, Malaysia, Pakistan, Vietnam, Philippines, and Bangladesh has increased from 2801.1 Mtoe in 2006 to 4340.6 Mtoe by 2015 reflecting more than 50% growth in ten years. However, the change in primary energy consumption during the same period in other regions is notably lower. North America witnessed a growth of −1.1%, South and Central America together witnessed 24.37% increase, Europe and Eurasia together witnessed a negative growth of −6.2%, Middle East reflected 49.4%, and African region reflected 28.8% increase' (BP 2016) (Fig. 4.2).

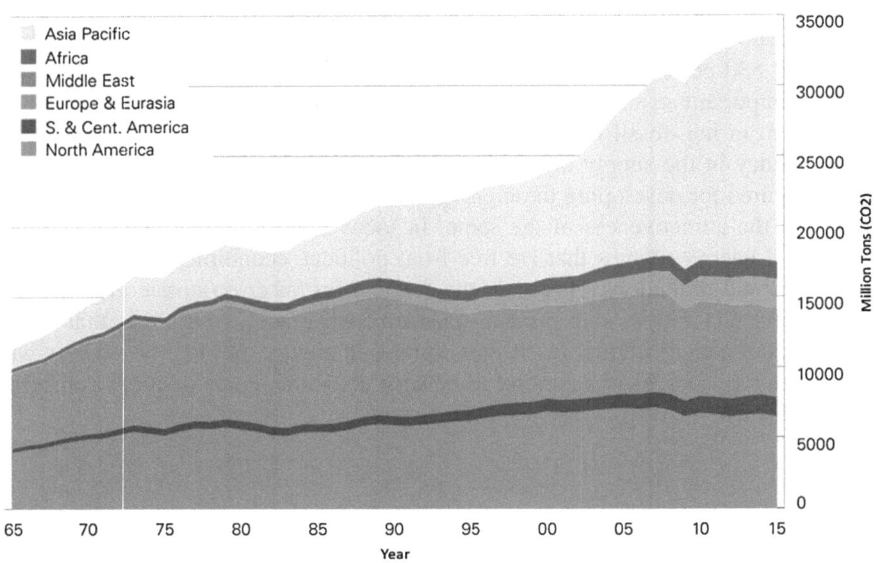

Fig. 4.1 Carbon dioxide emission. *Source* BP statistical review of world energy

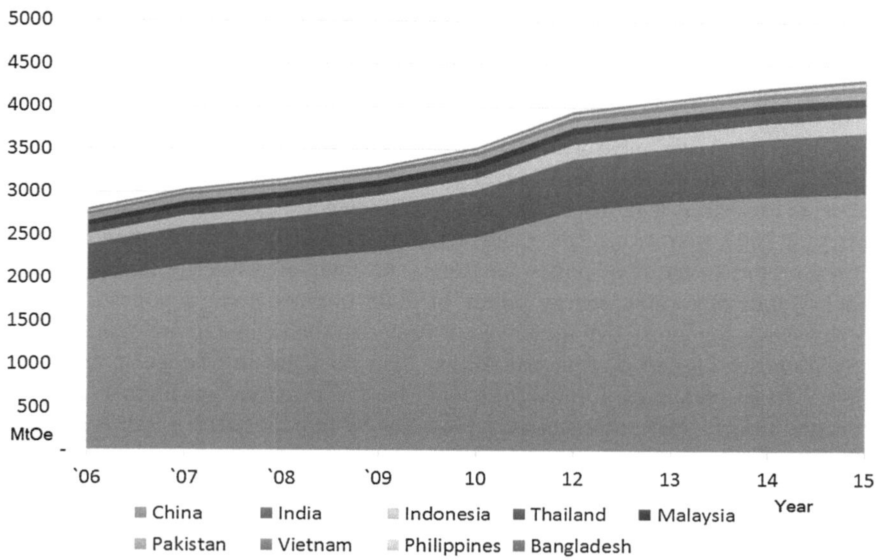

Fig. 4.2 Primary energy consumption in developing Asia. *Source* BP statistical review of world energy

Making of 'climate mitigation policy' gained momentum in developing countries after the 2007 'Bali Action Plan' adopted during the 13th Conference of Parties' (CoP 13) meeting held under the leadership of United Nations Framework Convention on Climate Change (UNFCCC). As part of the decision, developing countries too have agreed to put forward tangible actions to reduce the adverse effects of climate change impacts through Nationally Appropriate Mitigation Actions (NAMAs). Under NAMAs, developing countries have the flexibility in identifying appropriate measures that would minimise the impact of climate change. Unlike the developed economies that are required to take time bound emission reduction steps, NAMAs give more flexibility to developing economies as they are voluntary measures. This was noted as one of the major initiatives from the developing economies. After the 15th Conference of Parties held in Copenhagen in 2009, developing countries agreed to communicate their emission reduction strategies depending on their respective national sociopolitical and economic circumstances. The strategies were meant to address emission reduction concerns to be achieved before 2012. While certain countries proposed to reduce emission intensity of GDP, few other countries planned to make absolute reduction by devising ways to cap and minimise emissions. China proposed to reduce emission intensity of carbon dioxide by 40–45% by 2020 from that of 2005 levels, while India's target was to cut down CO_2 emission intensity by 20–25% by 2020 from that of 2005 levels. On the other hand, Indonesia proposed 26% of absolute carbon emission reduction by 2020 from that of 2005 levels. In order to achieve this target, Asian countries have devised various plans that would eventually help reduce energy-related emissions. The immediate target areas for energy reduction policies were industry, transportation, and household. Many of these countries have proposed a basket of policies that would aim at low-carbon development.

In the post-2009 period, India has proposed several policies that were in tune with other domestic measures formed earlier in the decade. Specific initiatives aimed at inclusive low-carbon development in India constituted several policy elements as well as legal tools. The Energy Efficiency Act of 2001 and Electricity Policy of 2003 were two of the major legal instrument for strengthening domestic energy sector. In terms of policy measures, the Integrated Energy Policy of 2006 and the draft renewable energy policy of 2005 outlined India's policies towards energy sector for future. All these were reflected in the eight national missions that were proposed as part of National Action Plan for Climate Change (NAPCC) of 2008. These initiatives formed a strong base for achieving India's emission reduction targets. Newer targets were proposed by India in 2015 which is called as Intended Nationally Determined Contribution (INDC) under which it has made strategies to reduce emissions intensity of GDP by 33–35% by 2030 below the 2005 levels.

With regard to China, the policy document on 'China's Policies and Actions for Addressing Climate Change' by the National Development and Reform Commission outlined its actions to guide the low-carbon development plan (NDRC 2009). The policy document which was based on several economic indicators from within the country had laid out plans for future. As per the target, China planned to reduce emission intensity of its GDP by 40–45% by 2020 from that of the 2005 levels. While this plan highlighted actions up to 2020, a detailed and more elaborate plan was made by China under the Intended Nationally Determined commitments (INDC) in 2015 which highlighted its actions up to 2030. This presented revised targets, 'including the target to peak CO_2 emissions by 2030 at the latest, lower the carbon intensity of GDP by 60–65% below 2005 levels by 2030, increase the share of non-fossil energy carriers of the total primary energy supply to around 20% by that time, and increase its forest stock volume by 4.5 billion cubic metres, compared to 2005 levels' (CAT 2015).

As indicated in Table 4.1, countries such as China, India, Thailand, Indonesia, and Philippines have significant emission reduction objectives laid out in the communication to the UNFCCC. However, the trend in currently planned energy mix in these countries will pose a challenging task in meeting the emission reduction objectives. Greater share of non-fossil fuel sources will be a strategic necessity.

4.4 Changing Energy Landscape of Asian Economies

Power sector is responsible for the largest chunk of global carbon emissions, primarily due to its dependence on coal and gas. According to OECD, 'the electricity sector is still far from being low-carbon as it continues to be dominated by fossil

Table 4.1 Emission reduction plans in developing Asia

Country	Copenhagen commitment (2020 targets)	INDC (2030 targets)
China	Reduction of CO_2 emissions intensity of GDP by 40–45% below 2005 levels by 2020	Reduction of CO_2 emissions intensity of GDP by 60–65% below 2005 levels by 2030
India	Reduction of CO_2 emissions intensity of GDP by 20–25% below 2005 levels by 2020	Reduction of CO_2 emissions intensity of GDP by 33–35% below 2005 levels by 2030
Indonesia	Emission reduction of 26% below business-as-usual (BAU) by 2020	Conditional 41% reduction below BAU by 2030 (with sufficient international support)
Thailand	Thailand proposes action in the energy and transportation sectors to reduce emissions between 7 and 20% below projections for 2020 (UNFCCC 2015a)	By 2030, reduce energy intensity (energy use per unit of GDP) to 25% below 2005 levels, across all economic sectors (IIP 2013)
Malaysia	Reduction of GHGs emissions up to 40% by 2020—Based on carbon intensity (UNCRD 2009)	Malaysia intends to reduce its greenhouse gas (GHG) emissions intensity of GDP by 45% by 2030 relative to the emissions intensity of GDP in 2005 (UNFCCC 2015b)
Pakistan	No specific targets on emission reduction	Reduce emissions by 10% by domestic actions (Dawn 2015)
Vietnam	No specific target on emission reduction. Showed support to emission reduction initiatives of UNFCCC based on CBDR principle	Reduce GHG emissions by 8% compared to BAU, and emission intensity per unit of GDP will be reduced by 20% compared to the 2010 levels (Huy 2010)
Philippines	No specific targets on emission reduction	The Philippines intends to reduce greenhouse gas (carbon dioxide equivalent) emissions by about 70% by 2030 relative to its business-as-usual estimate (UNFCCC 2015c)

Source http://www.ndf.fi/sites/ndf.fi/files/news_attach/3._monre_ghg_reduction_presentation.pdf

fuels, with 41% power generated at the global level by coal and 22% by gas' (NEA 2015). With growing demand for electricity, Asian economies are seen as major energy demand centres of future. 'Asia and the Pacific's primary energy demand is projected to increase at 2.1% per year over the outlook period (2010–2035)—faster than the projected world average growth rate of 1.5% per year during the same period. Correspondingly, primary energy demand of Asia and the Pacific will reach 8,358.3 million tons of oil equivalent (Mtoe) by 2035, which is up from 4,985.2 Mtoe as per 2010 estimates (ADB 2013a)'. The projected demand analysis also evinces that demand for oil, gas, and coal would increase tremendously. 'Demand for coal in Asia and the Pacific will increase by 52.8% from 2010 to 2035, reaching 3,516.3 million tons of oil equivalent (Mtoe) by 2035; oil demand in Asia and the

Pacific is projected to increase by 1.9% yearly over the outlook period and reach 1,973.0 Mtoe by 2035, 59.3% higher than the 1,238.2 Mtoe in 2010; and demand for natural gas is projected to increase at 3.9% per year, reaching 1,463.2 Mtoe in 2035, 2.6 times the 2010 level of 566.7 Mtoe (ADB 2013a)'.

In view of the increasing demand for electricity, investment is increasing in Asia. However, most of the investment is towards increasing power generation facilities that would depend on either coal or gas. As per the International Energy Agency's analysis on the cumulative investment in energy supply and energy efficiency during the period 2014–2035, Asia and Oceania together will account for approximately $1901 million worth investment. The IEA further states that, approximately $1157 million will be invested in the electricity generation sector, mostly based on thermal facilities (IEA 2014a, b, c)'. The huge demand for fossil fuels for electricity generation highlights that the hazards of emission will increase manifold.

4.5 Nuclear Power in Asia's Low-Carbon Energy Mix

Three countries in developing Asia, namely India, China, and Pakistan have (see Table 4.2) operable nuclear facilities currently. With 35 operational and 20 under construction, China leads the developing Asian region in terms of nuclear energy generation. It has plans to significantly increase the number of nuclear power facilities in the coming years. With 21 operational and six under construction, India stands second to China in terms of nuclear power generation. Pakistan has relatively smaller number of nuclear reactors and has been largely supported by the Chinese nuclear industry (Braun 2008; Hibbs 2014). So far, the nuclear industry in developing Asia has limited national presence.

Table 4.2 Operable nuclear facilities in developing Asia

Country	BWR	FBR	HTGR	PHWR	PWR	Total reactors
China		1	1	2	51	55
Operational		1		2	32	35
Under construction			1		19	20
India	2	1		22	2	27
Operational	2			18	1	21
Under construction		1		4	1	6
Pakistan				1	5	6
Operational				1	2	3
Under construction					3	3
Gross electrical capacity	320	525	211	8816	58,007	67,879

BWR Boiling water reactor, *FBR* Fast breeder reactor, *HTGR* The high-temperature gas-cooled reactor, *PHWR* Pressurised heavy water reactor, *PWR* Pressurised water reactor
Taiwan not included as part of China in the estimates
Source Nuclear power reactor information system of International Atomic Energy Agency, 2016

The most important and noticeable policy change was that the national energy policies in many countries began to undergo notable transitions to accommodate higher share of non-fossil fuels. The process of transition initiated in many countries included both supply side as well as demand side energy management measures aimed at achieving a newer mix of cleaner sources and technologies.

Analysis by ADB points that a possible increase of 202% in energy consumption between the period 2010 and 2035 will mainly be met by fossil fuels. If no specific policy changes are done to improve the energy mix, the region's energy consumption will continue to be dominated by coal and other fossil fuels and that 'local air and water quality, water availability, land use, and global climate all stand to suffer greatly if projected energy demand is met mostly by fossil fuels' (ADB 2013b). The region already suffering from the toxic emissions sets to face severe challenges in the years to come. The energy and transport sectors generate 70% of nitrogen oxide emissions and 80% of emissions of sulphur dioxide and particulate matter (PM10, referring to particles that are less than 10 μ in diameter and therefore able to penetrate deep into the respiratory tract) (ADB 2013a, b) (Fig. 4.3).

It is observed that with the increasing trend in emissions, Asia will be responsible for more than half of the total emissions in the world. Analysis done by International Energy Agency has made clear that to cut down emissions an advanced policy pathway needs to be adopted by the Asian region with an aim of addressing growing emissions. While the current policy scenario can pose challenges, the 450 scenario (under which IEA recommended policies that need to be

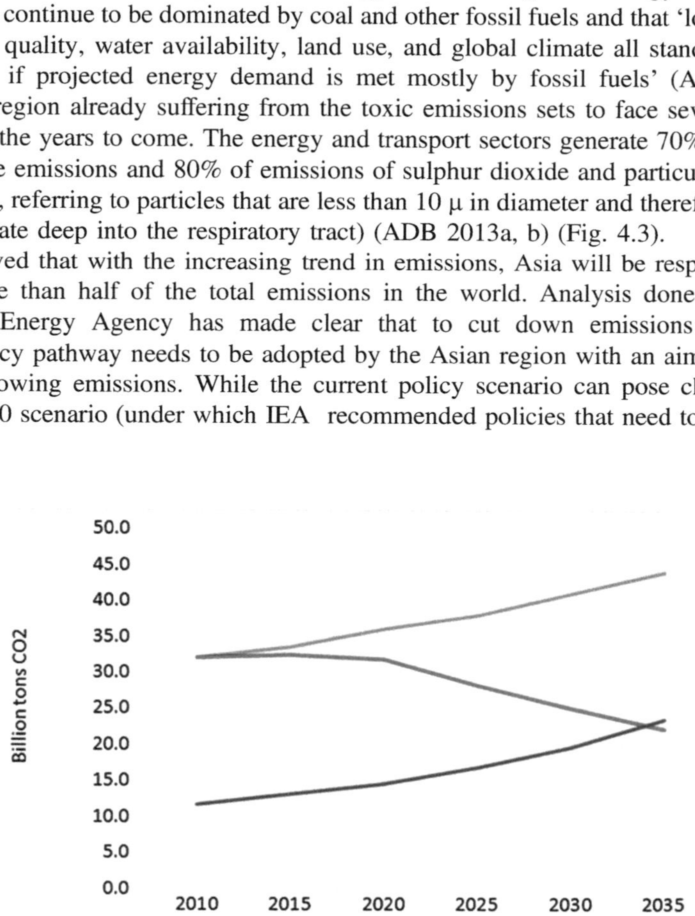

	2010	2015	2020	2025	2030	2035
—IEA current policies scenario (world)	32.0	33.5	36.0	38.0	41.0	44.0
—IEA 450 scenario (world)	32.0	32.4	31.8	28.2	25.0	22.1
—Developing Asia carbon dioxide emissions	11.7	13.1	14.5	16.9	19.6	23.5

Fig. 4.3 Projected energy-related carbon dioxide emissions from developing Asia (Billion tons). *Source* http://www.adb.org/publications/asian-development-outlook-2013-asias-energy-challenge

adopted by the countries across the world to limit the global emissions below 450 ppm to keep the global temperature rise below 20 °C by 2020) can be better policy alternatives for countries. Based on the current energy consumption pattern in Asian region, the emissions will be above half of the global emission level as mentioned earlier.

It is estimated that the 'electricity demand in Asia and the Pacific is projected to be more than double between the period 2010 and 2035, increasing from 7,010.4 TWh in 2010 to 16,169.2 TWh in 2035. Among developing members, this will grow slightly faster, at an annual rate of 3.8% on average through 2035, and increase their share in the total electricity demand of Asia and the Pacific from 82.3% in 2010 to 91.1% in 2035' (ADB 2013a). China and India are expected to have highest demand in the region. Ensuring continuous energy supply for these countries require not only huge investment in the electricity generation plans in non-conventional areas but also points out the need for transitioning to new energy mix, where maximum utilisation of sources such as nuclear, solar, wind, and biomass is ensured.

From the climate perspective, nuclear power is seen as a clean energy and has the potential to offer cleaner and cheaper energy supply to developing world. The Intergovernmental Panel on Climate Change (IPCC) while recommending the need for switching to clean energy resources has highlighted that nuclear energy has been one of the cleaner sources. The report of the Working Group III on Mitigation (in 2001) highlights that 'Nuclear energy could replace baseload fossil fuel electricity generation in many parts of the world if acceptable responses can be found to concerns over reactor safety, radioactive waste transport, waste disposal, and proliferation' (IPCC 2001). In the synthesis report on climate change released in 2007 IPCC reiterates that 'for lower stabilisation levels, scenarios (there is a need to) put more emphasis on the use of low-carbon energy sources, such as renewable energy, nuclear power and the use of CO_2 capture and storage (CCS)' (IPCC 2007) (Fig. 4.4).

GHG emissions for various fuel types show that nuclear is at par with hydroelectric of wind energy. As noted by observers, 'to keep the global temperature rise under 3 °C will require an increase in non-carbon electricity generation from 34% (nuclear plus hydro) now to 48–53% by 2030, along with other measures. With a doubling of overall electricity demand by then, and a carbon emission cost of US$50 per tonne of CO_2, nuclear power's share of electricity generation is projected by IPCC to grow from 16% now to 18% of the increased demand' (WNA 2016a, b, c, d, e).

In order to meet the energy demand, many in the developing Asian region also have planned to increase the cleaner energy share in the years to come. In China as the government has committed to cut down the emission intensity of GDP to 60–65% in 2030 from 2005 levels, greater thrust has been given on increasing clean energy including nuclear. The 13th Five Year Plan of China which was finalised in March 2016 included several nuclear projects too taking into consideration the countries upcoming demand for cleaner energy. In the case of India, the surging demand for electricity and use of cleaner sources due to the climate concerns is making way for greater investment in nuclear power. Though the antinuclear public sentiment has been a major challenge for the nuclear energy sector, meeting the emission intensity targets of 33–35% by 2030 will demand nuclear power.

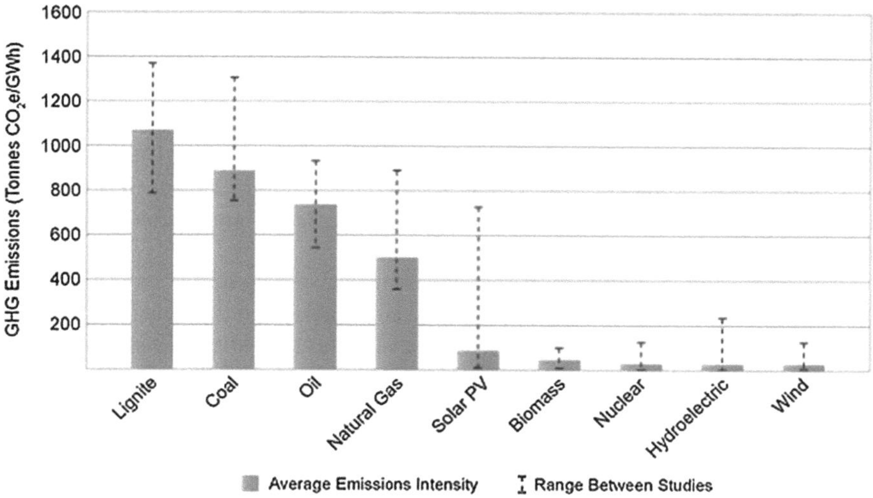

Fig. 4.4 GHG emissions intensity. *Source* Nuclear is recognised as a crucial mitigation technology, and many countries are planning to use nuclear generation to meet their emission reduction and energy supply objectives (WNA 2016)

In Indonesia, National Energy Policy, which was made in 2014, updated the country's targets, an increase of renewable energy to 23% of primary energy supply by 2025 (Climateactiontracker 2016). The country has no nuclear reactors as of now but has made serious proposals to build nuclear facilities in the coming years. In April 2015, the 'Rosatom—Russian nuclear company along with the Indonesian companies led by NUKEM Technologies—won contracts for a preliminary design for the multi-purpose 10 MWe HTR reactor in Indonesia. Apart from this, Martingale—a company from USA along with its consortium members including Indonesian companies—agreed to build a ThorCon thorium molten salt reactor by 2021' (WNA 2016a, b, c, d, e). Among the developing economies in the Asian region with new nuclear new build plans, Vietnam has one of the well-developed nuclear energy plans. With the support of Russia, South Korea, and Japan, Vietnam has proposed several nuclear reactor plans. With the currently projected power demand scenario, from 2020 onwards, the country needs to increase the share of nuclear from 1.5 to 6% in 2025 then to 8% in 2030 and eventually to 20–25% by 2050. Nuclear power produced about 4332.70 GWh of electricity in 2015, which was roughly about 4% of total electricity generated in the country. The country has relatively smaller programme than its two neighbours but has been getting enormous support from Beijing. In case of Thailand, nuclear power does not play any role in the country's energy mix as of now. Similarly, countries such as Malaysia and Philippines too do not depend on nuclear power but have nuclear power development plans for the coming years.

4.6 Nuclear Power Development in Asia: Opportunities and Challenges

Several points are in public debate with regard to the suitability of nuclear as a key element in the region's energy mix. First, the concern about climate change and the role of nuclear a promising component, second, the economics of nuclear facilities, third fuel availability, security, and safety concerns, and fourth public perceptions towards nuclear.

The cost forms a significant parameter in nuclear policy decisions in the world. Though the capital cost of nuclear power is remarkably high than the other sources, low operation and maintenance ensures greater acceptability of nuclear power. 'Investment typically represents more than 50% of total generation cost of nuclear electricity. Once built, on the other hand, nuclear plants have rather low fuel and operating costs, compensating for the need to amortise their capital investment. Since the cost of uranium represents only 5% of the cost of electricity from nuclear plants, even a significant rise in the cost of uranium would have little impact on the total cost of nuclear-generated electricity' (Bertel and Morrison 2001). Similarly, the availability of uranium for nuclear power generation in the region often surfaces in energy security debate. It is expected that unlike the conventional fossil fuels deposits that may already be declining due to continued exploration and production over centuries, fuel sources for nuclear will last for longer period. Nuclear Energy Institute, while quoting the 2009 report of Nuclear Energy Agency (OECD), observes that uranium resources are adequate to meet nuclear energy needs for at least the next 100 years at present consumption levels and more efficient fast reactors could extend that period to more than 2,500 years (NEI 2016a, b). An estimate of the global distribution of identified uranium resources points that Kazakhstan, Uzbekistan, Mongolia, China, and Australia have uranium deposits (which is feasible to be explored at $130/kg Uranium). While Australia tops in uranium deposits with 29% of world resources, Kazakhstan has 12%, China has 3%, and Uzbekistan as well as Mongolia has 2% each of the global deposits (NEA 2014). Though the distribution of these deposits is uneven in the world, with a prevailing well-defined uranium trade structure, it can be assumed that countries which have nuclear power facilities will continue to have access to nuclear fuels and lack of domestic availability of fuels are unlikely to create any supply challenges. However, countries like India which do not have major uranium deposits will continue to depend on international trade. Often, this has been raised by experts to question the feasibility of nuclear power generation in countries which do not have recoverable uranium deposits. As the trade of nuclear fuel and materials are monitored by the international agencies, it is important for the consuming countries to adhere to various international norms. This also is one of the hurdles as a country considers dependency on nuclear. Though India does not have nuclear major nuclear fuel deposits, it has made significant strides in the nuclear technology over the past many years. It has also been relying on nuclear power for decades and has developed strong relation with major nuclear sector players. Moreover, India has a strong commitment to the international norms and also has exceptional track record of not involving in any proliferation of nuclear materials. However, with regard to other countries in the region, the feasibility of nuclear energy has always come under question due to the concerns about security threat and adverse public perception.

Nuclear power also faces several critical challenges, both domestic and external. While external challenges are mostly related to the limitation in availability of technology and fuel from outside supply sources, domestic challenges pose more severe threat to nuclear sector's development in any economy. These include 'lack of enough funding', 'security challenges', 'safety concerns', 'adverse public perception' etc. that seriously affect the development of nuclear power industry. An orbit model is used here to represent the challenges to nuclear energy development in a country and the potential domestic and external challenges. In this model, two elliptical orbits are used—inner orbit and outer orbit—to represent the domestic and external challenges, respectively. The elliptical orbits reflect that challenges to nuclear sector are dynamic and would keep changing. The distance between the indicative threat point in the orbit and the centre can be taken as measure of impact —closer the threat, higher the impact. It also argues that the domestic challenges despite emerging from within the borders of a country often pose bigger threat to the nuclear sector development than the external issues (Fig. 4.5).

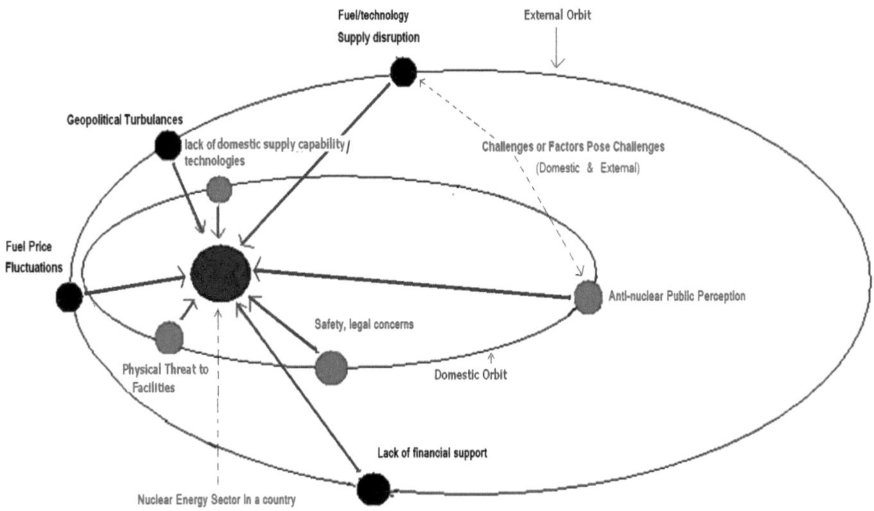

Fig. 4.5 Orbits of energy security: domestic and external challenges. *Source* Domestic challenges often appear more critical than external challenges to the nuclear energy sector. While the extent to which this argument can be validated may vary country to country, nuclear sector which depends a conducive domestic environment appears to be highly influenced by domestic challenges in any country (Janardhanan 2008). This is for indicative purpose to explain how challenges to nuclear energy sector can be dynamic in a country or in a given geographic context

In many countries, one of the biggest challenges to the development of nuclear power is the adverse public sentiments. In the post-Chernobyl and post-Fukushima period, the intensified debates on safety and security of nuclear facilities across the world witnessed a surge in antinuclear sentiments. The fear of nuclear power technology was undeniably backed by political and vested interests. In particular, after the Chernobyl accident in Ukraine, the debates in many countries encountered high polarisation among political sections based on the nuclear energy policy. Notably, the political factions have apparently used this debate to intensify their attack against the government. While radiation effects have undeniable adverse impacts on human beings and environment, the background radiation that exists in our surroundings is conveniently ignored in these debates. In many other occasions, human beings are exposed to much higher level of radiation than often claimed to be affected by the nuclear facilities. This includes utilisation of different machinery, use of granite rocks, or even airline travel which presents much severe radiation exposure to human and other living environment. In the post-Fukushima period, the antinuclear sentiments that happened in many countries in Asian region also can be seen as intertwined with political opportunism and anti-government sentiments.

The concerns regarding security of nuclear facilities demand highest attention in this regard. Keeping nuclear materials and facilities out of reach for potential non-state threats is one of the biggest responsibilities for a state. The fact that no single country is completely safe from any potential security threat demands the state to provide high security environment. This inevitably boosts the cost of maintaining a nuclear facility as well as for the nuclear power generation. 'One of the most serious threats facing the USA and other countries is the possibility that other nations and terrorist organisations could steal a nuclear warhead or nuclear weapon-usable materials from poorly secured stockpiles around the world, or that nations could divert nuclear materials intended for peaceful purposes to the development of nuclear weapons' (Trimble 2015).

The antinuclear public perception is also strengthened by the arguments which equate the nuclear waste storage and reprocessing to the dangers of nuclear proliferation. One can observe that no other energy sector and energy technology is as misinterpreted and misconceived as nuclear energy. The vested interest groups have always used the imagery of the atomic bomb attacks on Hiroshima and Nagasaki and have successfully misled the global community about the real pros and cons of nuclear power.

4.7 Conclusion

In the post-Fukushima period, it has been observed that nuclear power sector in the Asian region may wane due to its vulnerabilities to withstand any safety or security challenge. The 'false' concerns about the vulnerability of nuclear facilities are often over-projected. The number of human casualties per unit of different sources of electricity shows that nuclear has been responsible for the lowest so far. While the impact of a major (INES 7th scale[2]) accident may cause damage to the human as well as the environment, chances of such events are extremely rare with the current level of technological maturity. The ongoing climate debate and the demand for cleaner fuels together will contribute to the resurgence of nuclear power sector rather faster. Though in the short term and medium term the nuclear energy industry will not witness any significant growth, the long term offers potential opportunities for nuclear power sector.

While nuclear energy would be respected for its value as long as a long-term energy source, the development of the same in the Asian region can encounter many hurdles. The biggest challenge to nuclear energy development in the Asian region would be from the sociopolitical concerns that often snowball into antinuclear public sentiments. Besides, the often projected arguments that alternative energy sources are more efficient and environment friendly can also limit nuclear sector's chances of winning any public debate easily. Any success of nuclear power debate in convincing the society can happen only through adequate education with regard to the real advantages and disadvantages of nuclear. This should be provided to the society in a transparent and impartial manner by both governments and other nuclear establishments in the region.

[2]According to IAEA, The International Nuclear and Radiological Event Scale (INES) uses a numerical rating to explain the significance of events associated with sources of ionizing radiation. Events are rated at seven levels: Levels 1–3 are 'incidents' and Levels 4–7 'accidents'. The scale is designed such that the severity of an event is approximately ten times greater for each increase in level of the scale. These levels consider three areas of impact: people and the environment, radiological barriers and control, defence in depth. Events without safety significance are rated as Below Scale/Level 0. Events that have no safety relevance with respect to radiation or nuclear safety are not rated on the scale. The events are rated as 0 = No safety significance, 1 = Anomaly, 2 = Small incident, 3 = Serious incident, 4 = Accident with local consequences, 5 = Accident with wider consequences, 6 = Serious accident, 7 = Major accident involving human casualties and environment damage (IAEA 2016).

References

ADB. (2013a). *Energy outlook for Asia and Pacific*. Manila: Asian Development Bank.

ADB. (2013b). *Asian development outlook*. Manila: Asian Development Outlook.

ADB. (2014). *Enegy security, sustainability and affordability in Asia and the Pacific*. Available at https://www.adb.org/sites/default/files/publication/42782/ewp-401.pdf.

Armstrong, I. (2015). *Russia is creating a global nuclear power empire*. Available at http://globalriskinsights.com/2015/10/russia-is-creating-a-global-nuclear-power-empire/. Accessed 1 May 2016.

Bertel, E., & Morrison, R. (2001). Nuclear energy economics in a sustainable development perspective. *NEA News, 19*(1), 14–17.

BP. (2016). *Statistical review of world energy*. London: British Petroleum.

Braun, C. (2008). Security issues related to Pakistan's future nuclear power program. In: SSI (ed.), *Pakistan's nuclear future: Worries beyond war* (pp. 277–346). Carlisle: Strategic Studies Institute.

BusinessLine. (2015). *India has cut poverty by half, but progress is uneven: UN*. Available at: http://www.thehindubusinessline.com/economy/india-has-cut-poverty-by-half-but-progress-is-uneven-un/article7395907.ece. Accessed 17 May 2016.

CAT. (2015). *Climate action tracker*. Available at http://climateactiontracker.org/countries/china.html. Accessed 12 July 2016.

Chu, S. (2010). *Wall Street Journal*. Available at: http://www.wsj.com/articles/SB1000142405274870423130457509213023999278. Accessed 6 Sep 2016.

Climateactiontracker. (2016). *Climate action tracker*. Available at http://climateactiontracker.org/countries/indonesia.html. Accessed 3 July 2016.

Dawn. (2015). *Pakistan's national climate plan—neither 'ambitious' nor 'imaginative'*. Available at http://www.dawn.com/news/1210650. Accessed 20 May 2016.

EIA. (2015). *Energy information administration*. Available at http://www.eia.gov/pressroom/presentations/sieminski_09212015.pdf. Accessed 22 Feb 2016.

Hibbs, M. (2014). Power loop: China provides nuclear reactors to Pakistan. In: IHS (ed.), *Jane's intelligence review*, pp. 50–53. New York: Jane's Information Group.

Huy, L. Q. (2010). *Nordic development fund*. Available at http://www.ndf.fi/sites/ndf.fi/files/news_attach/3._monre_ghg_reduction_presentation.pdf. Accessed 1 Feb 2016.

IAEA. (2014). *Advances in small modular reactor technology developments*. Available at https://www.iaea.org/NuclearPower/Downloadable/SMR/files/IAEA_SMR_Booklet_2014.pdf.

IAEA. (2015). *Climate change and nuclear power*. Vienna: International Atomic Energy Agency.

IAEA. (2016). *The international nuclear and radiological event scale*. Available at http://www-ns.iaea.org/tech-areas/emergency/ines.asp. Accessed 1 July 2016.

IEA. (2014a). *Energy technology perspectives—Harnessing electricity's potential*. Paris: International Energy Agency.

IEA. (2014b). *World energy investment outlook*. Paris: International Energy Agency.

IEA. (2014c). *World energy outlook*. Paris: International Energy Agency.

IEA. (2015a). *Climate change and nuclear energy*. Paris: International Energy Agency.

IEA. (2015b). *World energy outlook*. Paris: International Energy Agency.

IEA. (2015c). *World energy outlook*. Paris: International Energy Agency.

IEA. (2016). *Nuclear energy*. Available at https://www.iea.org/about/faqs/nuclear/. Accessed 1 May 2016.

IIP. (2013). *Thailand's GHG emissions reductions policies*. Washington: Institute of Industrial Productivity.

IPCC. (2001). *Report of working group on climate mitigation*. Geneva: Intergovernmental Panel on Climate Change.

IPCC. (2007). *Synthesis report of IPCC assessment report-4*. Geneva: Intergovernmental Panel on Climate Change.

IPCC. (2014). *Fifth assessment report—Summary for policy makers*. Available at https://www.ipcc.ch/pdf/assessment-report/ar5/wg3/ipcc_wg3_ar5_summary-for-policymakers.pdf. Accessed 20 Aug 2016.

Janardhanan, N. (2008). *India and nuclear power: Examining socio-political challenges to energy security*. London: World Nuclear Association.

King, D., et al. (2015). *Climate change—A risk assessment*. Mumbai: Foreign Commonwealth Office, UK.

NDRC. (2009). *Department of Climate Change, National Developmentand Reform Commission*. Available at: http://www.ccchina.gov.cn/WebSite/CCChina/UpFile/File571.pdf. Accessed 21 July 2016.

NEA. (2014). *Uranium 2014: Resources, Production and Demand*. Paris: NEA-IAEA.

NEA. (2015). *Nuclear energy: combating climate change*. Paris: OECD.

NEI. (2016a). *Nuclear Energy Agency*. Available at http://www.nei.org/Issues-Policy/Nuclear-Fuel-Supply. Accessed 14 July 2016.

NEI. (2016b). *TVA files historic application for small modular reactors*. Available at http://www.nei.org/News-Media/News/News-Archives/TVA-Files-Historic-Application-for-Small-Modular-R.

Smil, V. (2004). War and energy. In C. J. Cleveland (Ed.), *Encyclopedia of Energy* (pp. 363–371). Winnipeg: Elsevier Inc.

Trimble, D. (2015). GAO: two decades evaluating the impact and effectiveness of U.S. nuclear and radiological material security programs. In S. Apikyan & D. Diamond (Eds.), *Nuclear threats and security challenges* (p. 284). Los Angeles: Springer.

UN. (2014). *Sustainable development platform*. Available at https://sustainabledevelopment.un.org/?page=view&nr=164&type=230&menu=2059. Accessed 21 Apr 2016.

UN. (2015a). *United Nations in India*. Available at http://in.one.un.org/img/uploads/SDGs_PressRelease_September_2015.pdf. Accessed 22 April 2016.

UN. (2015b). *Sustainable development goals*. Available at http://www.un.org/sustainabledevelopment/health/. Accessed 01 Apr 2016.

UN. (2015c). *United Nations in India*. Available at http://in.one.un.org/page/first-national-consultation-on-sdgs-in-india. Accessed 20 Apr 2016.

UN. (2016d). *Press release: Access to technology key to address climate change and development under UN's SDGs*. New Delhi: United Nations in India.

UNCRD. (2009). *United Nations Centre for Regional Development (UNCRD)*. Available at http://www.uncrd.or.jp/content/documents/PS2_07_Nadzri_MHLG.pdf. Accessed 23 June 2016.

UNFCCC. (1997). *Kyoto Protocol*. Kyoto: United nations framework convention on climate change.

UNFCCC. (2015a). *Climate news room*. Available at http://newsroom.unfccc.int/unfccc-newsroom/thailand-submits-nama/. Accessed 10 July 2016.

UNFCCC. (2015b). *Intended nationally determined contribution of the government*. Available at http://www4.unfccc.int/Submissions/INDC/Published%20Documents/Malaysia/1/INDC%20Malaysia%20Final%2027%20November%202015%20Revised%20Final%20UNFCCC.pdf. Accessed 15 June 2016.

UNFCCC. (2015c). *Intended nationally determined contributions*. Available at http://www4.unfccc.int/submissions/INDC/Published%20Documents/Philippines/1/Philippines%20-%20Final%20INDC%20submission.pdf. Accessed 3 June 2016.

WNA. (2015). *World energy needs and nuclear power*. Available at http://www.world-nuclear.org/information-library/current-and-future-generation/world-energy-needs-and-nuclear-power.aspx. Accessed 5 Jan 2016.

WNA. (2016a). *Asia's nuclear energy growth*. Available at http://www.world-nuclear.org/information-library/country-profiles/others/asias-nuclear-energy-growth.aspx.

WNA. (2016b). *Climate change—The science*. Available at http://www.world-nuclear.org/information-library/energy-and-the-environment/climate-change-the-science.aspx. Accessed 1 July 2016.

WNA. (2016c). *Emerging nuclear energy countries.* Available at http://www.world-nuclear.org/information-library/country-profiles/others/emerging-nuclear-energy-countries.aspx. Accessed 20 July 2016.

WNA. (2016d). *Nuclear energy and climate change.* Available at http://www.world-nuclear.org/focus/climate-change-and-nuclear-energy/nuclear-energy-and-climate-change.aspx. Accessed 1 July 2016.

WNA. (2016e). *Nuclear power in Indonesia.* Available at http://www.world-nuclear.org/information-library/country-profiles/countries-g-n/indonesia.aspx. Accessed 15 July 2016.

Author Biography

Dr. Nandakumar Janardhanan teaches in the Energy Studies Programme of Jawaharlal Nehru University (JNU), New Delhi, India. He specialises in energy and climate policy. The areas of his work include nuclear power, energy security and geopolitics, climate policy, low-carbon development and renewable energy. He has worked with several policy think tanks and research organizations of international repute. Dr. Janardhanan was a Fellow with the 'Climate CoLab' of Massachusetts Institute of Technology (2014–2015) and one of the lecturing staff for the Nuclear Energy training conducted by the International Atomic Energy Agency (2011–2014). He has been associated with the Institute of Chinese Studies (India) as Adjunct Fellow, as Member of Teaching and Advisory team at Strathclyde Centre for Environmental Law and Governance of the University of Strathclyde (Scotland) and as Fellow with the Institute for Global Environmental Strategies (Japan).

Chapter 5
Small Modular Reactors—Does Asia Need This Technology Overhaul?

Karthik Ganesan

5.1 Background to SMR

The acronym SMR refers to *small, modular reactors,* where *small* implies that the electrical output is less than 300 MWe.[1] '*Modular* means that these reactors can be assembled from prefabricated modules, with each module representing and constituting a portion of the final plant, rather than being built on site like conventional nuclear power plants' (Sovacool and Ramana 2014).

Small modular reactor (SMR) are being developed by different countries which have a large and mature civilian nuclear program and are looking at innovative technologies to put an end to the stagnation that the industry has seen in these countries. 'The development of small reactors began in the early 1950s for naval propulsion' (Vujic et al. 2012; Ingersoll 2015). Eventually, reactor designs and builds gyrated towards large sizes for various reasons—ranging from economies of scale to limited availability of sites—which then required maximum capacity installations at one site. Current SMR efforts attempt to simplify the design aspects of a nuclear reactor, provide modularity, passive safety features, enable faster construction and potentially reduce financial burden that has been associated with conventional nuclear reactors (Ingersoll 2015). Their deployment strategy allows for an incremental approach, thus reducing upfront capital requirements and enabling their use in geographies with limited grid capacity. It is also attractive to

This chapter is influenced by the author's work which he has conducted as part of his research at the National University of Singapore. The use of any content from, referenced appropriately, is fully permissible and I consent to any use of content from my thesis.

[1]Small, as defined by the IAEA, in the context of small and medium reactors (also SMR).

K. Ganesan (✉)
Council on Energy, Environment and Water, New Delhi, India
e-mail: karthik.ganesan@ceew.in

N. Janardhanan et al. (eds.), *Resurgence of Nuclear Power*,
https://doi.org/10.1007/978-981-10-5029-9_5

countries that do not currently have a nuclear program (or only a test or laboratory reactor), but are interested in kick-starting a program outside the laboratory. The limited evidence on SMR costs suggests that despite higher specific capital costs (on a per MW basis), as compared to large-scale reactors, they could provide economic benefits specifically on account of the phased financial outlay for modular units, the quicker commissioning of units, ability to co-site in facilities that already have necessary approvals and the possibility of learning and scale effects. 'In addition, some of the proponents of SMRs also indicate that co-generation and advanced process heat applications are other areas where these reactors will find use' (Stanculescu 2013).

In a widely read piece in the Wall Street Journal (2010), the then Secretary of Energy (US) Steven Chu suggested that the SMR design is based on the established light-water reactor technologies and would be deployed in (about) 10 years (Chu 2010). In the intervening years since Secretary Chu's piece, while much interest, including the USD 200 million grant from DoE to the SMR industry has been drummed up, it was only in May 2016 that the first 'Early Site Permit' for an exclusive SMR site was filed by the Tennessee Valley Authority with the US nuclear regulator (NRC). There has been no further progress on this first of a kind implementation of SMRs, and the timelines for application of a licence and start of construction are still not clear (NEI 2016). The IAEA, in an update on advancements in SMR technology development, states that 'projected timelines of readiness for deployment of the technology generally range from the present to 2025–2030' (IAEA 2014).

It is this sort of enthusiasm shown in Chu's piece that detractors particularly target. Sovacool and Ramana (2014) liken the vision of SMR proponents to fantasies and suggest that those with a vested interest in the nuclear industry (scientists and technology companies alike) are drumming up support for SMRs by appealing to a shared group consciousness, replete with elements of fantasy and notions that social constructs endorse. Some of these visions, albeit with a less harsh lens, will be explored in the subsequent section of this chapter and anchoring these in the developmental needs of Asia.

5.2 The Challenges and Opportunities in Pursuing SMRs in Asia

Asia is a hotbed of activity for nuclear power capacity addition, as compared to other parts of the world. Currently, there are 128 reactors that are operational, while a further 40 reactors are under construction, and nearly 89 are in the planning phase (WNA 2016). The appetite for the region to expand nuclear power capacity exists but the technologies being pursued are large LWRs and PHWRs. With the exception of China, the challenges associated with expanding nuclear power capacity are similar across the region. Disapproval or resistance from sections of the public, and delays as a result of licensing and land acquisition, and more stringent regulatory requirements (in the aftermath of Fukushima) are the some of the key barriers in Asia. With this background, it is useful to identify those gaps that the SMR technology/designs can help fill. We can understand the policy drivers that promote and deter nuclear energy (specifically SMRs) in developing and developed Asia through the following narrative.

5.2.1 Energy Access for Developing Asia

In any debate on sources of energy for the rapidly growing economies of developing Asia, the oft-repeated statement is that Asia needs any energy source it can get its hands on and cannot rule out contributions from any. RE, coal, oil, natural gas, hydro and nuclear—all are important. Unlike the West, where substitution of conventional power sources (and even nuclear) for RE is already underway, demand is still on the rise in Asia. Substitution of one source for another is still some time away.

The possibilities for technologies that can cater to the needs of communities spread far and wide are immense in India. As per the Census, nearly a third of the population reported using inferior sources of energy such as kerosene for lighting purposes (Census 2011). The Rajeev Gandhi Grameen Vidyutikaran Yojana (RGGVY), the flagship rural electrification program, has been under implementation for over a decade now. However, 'despite the concerted effort in bringing the wires closer and an increase in centralised generation capacity, 300 million people do not have access to reliable, affordable and sufficient quantities of electricity' (Jain et al. 2015). The rhetoric has always suggested that the grid is the best way to deliver power to the underserved masses of India, but despite sustained efforts, millions remain without reliable electricity supply, even in densely populated areas.

About 740 million people live on more than 50,000 islands across the world, and many of these rely on imported fossil fuels for their energy needs (Walton 2016). In some of the Pacific island states (such as the Philippines), oil still contributes a significant share of electricity generation and the ability to run the grid to small inhabited islands is often a difficult proposition. 'Indonesia (also an archipelago), which sees much reliance on coal, has nearly 60 million people that do not have access to electricity' (PWC 2013). With few islands enjoying grid connections to the mainland or even to neighbouring larger islands due to the high cost of submarine transmission cables, island power supply is generally unstable and unreliable.

While renewable energy has fulfilled the needs of many rural and remote communities (and is particularly attractive to tropical islands), the business models to scale up the energy service provided by RE have not dented the energy access deficit enough. The efficiency of these systems is still poor and, the design of appliances and battery systems (to store the energy), is still very much in inception. In areas where RE has shown promise, the demand for higher levels of electricity supply, after an initial taste of the offering, is a near universal phenomenon. In rural areas, which have a fairly dense cluster of homes, a solution such as that offered by

SMR can compete with RE-based mini-/micro-grids. The energy density of SMRs is a lot higher and can generate a lot more power with a lot lower land footprint.

The other challenge that entrepreneurs and policy makers are wary of is the need to integrate distributed energy generation sources with the grid, if that becomes necessary. The technological challenges associated with variability in generation (of RE sources) imply that their integration with the grid could prove expensive. As mentioned earlier, solar is predominant during the day while the requirement (peak power or otherwise) is at night time or after sunset. Electrochemical storage technologies are an essential component in the integration of RE, and in being able to supply when the sun is not shining or the wind is not blowing. This adds a significant cost to the overall generation. In India for example, the cost of storage is (per unit of electricity supplied through the stored system) larger than the cost of generating that electricity from RE-based systems (Kuldeep et al. 2016).

India is in the midst of an effort to increase RE capacity to 175 GW[2] (by 2022). The transmission system capacity augmentation has lagged behind generation for nearly three decades, and it is evident that in integrating such a large quantum of RE, significant investments in transmission, in hitherto unconnected areas will be needed. 'Some of these are being executed at a very high cost (USD 7.58 billion) and many more corridors would be required' (POWERGRID 2012). It is this high cost of transmission that has resulted in many calling for a 'use where you produce' approach and a push for decentralised production and consumption of energy. More importantly, most centralised energy provision schemes focus only on electricity and not cooking or heating. This is a significant challenge for many rural areas in India. An SMR has the potential to play the role of 'Energy from a Box' and offers various combinations of heat, power and cooling based on the requirements of the areas being served. Nearly, 75 million people in India live in villages which have more than 10,000 households. There are ~5000 villages such large villages in the country. These could offer a captive consumer base for a 50 MW unit to provide enough power (~4 kW per household, as is the case in urban households in India) to the households, to enjoy the same basket of services as urban India does today. If one takes productive need into consideration, these villages could easily consume as much power as 100 MW.

5.2.2 *Energy Security*

Though Asia is endowed with plentiful resources of fossil fuel resources, the distribution across geographic boundaries is skewed. West Asia and some parts of ASEAN have a rich resource base of natural gas and/or petroleum. As per the ADB, nearly 40% of China's energy and nearly 35% of India's total primary energy will be imported by

[2]The current total installed capacity in India is ~310 GW. This sets in context, the massive efforts required.

2035 (ADB 2014). It is also increasingly clear that in a globalised world with functioning markets for many energy commodities, most countries will not follow autarkic definitions of energy security and pursue energy independence. Energy security is also not just about getting access to the fuel but is a multidimensional issue. Energy security could be defined in a more complete manner as '*as the availability of adequate quantities of critical resources, at prices that are affordable and predictable, with minimum risk of supply disruptions, to ensure sustainability for the environment and future generations*' (Bery et al. 2016). From the lens of energy security, the case for pursuit of nuclear energy is stronger, as the sustained dependence on import of fuel is much lower than for other conventional sources of energy.

Though SMRs and the conventional NPP designs are estimated to have similar burn-up rates, a lower thermal efficiency implies the need for additional to produce the same amount of electricity. 'This would result in higher variable costs over the life of the plant and needs to be offset significantly through economies of scale in the manufacturing costs of nth of a kind units' (Pannier and Skoda 2014; American Nuclear Society 2010). While the quantum of fuel required is significantly lower than in fossil fuel-fired generators, SMRs will result in higher imports of fuel as well.

Even in the case of RE (specifically solar energy in today's world), the manufacture of important components (namely the cells in solar-PV) are restricted to a few countries that are able to manufacture at large enough scales to bring down the costs across the board. It is very likely that even in the case of SMRs, the capability to manufacture will remain in the hands of a few countries. Given that the modules come prefabricated, the opportunity to locally source material and inputs is also limited. The strict compliance and precision requirements also mean that developing countries without an active program will not be in any position to take on manufacturing of any of the components of a module. The energy security implication (at least the conventional view of energy security) is that small and developing countries will again find themselves dependent on outside entities for securing their energy, even with this technologically advanced solution. When viewed under the lens of improving energy security, SMRs do not necessarily address the issue sufficiently.

5.2.3 Sustainable Development and Climate Change

'Most studies by agencies such as the IEA suggest that coal will be the mainstay of power generation in most parts of Asia' (IEA 2015). A shift away to non-fossil energy sources is essential to cutback the growth of emissions. 'Achieving such stringent climate goals will require substantial reductions—in the order of 50% below current levels—in the emissions of greenhouse gases (GHG) by 2050 and deeper cuts beyond' (Iyer et al. 2014). 'Among the low carbon technologies, nuclear has one of the lowest life cycle emissions of GHG' (IAEA 2015). 'A phase-out of nuclear technologies would imply incremental mitigation costs ranging from 7 to 13% across various scenarios' (IPCC 2014). In fact, many scenarios suggest the 'need for negative emissions technologies such as CCS' (IEA 2014) will have to be rolled out in order to keep global warming within acceptable levels, agreed upon in successive UNFCC negotiations.

'Without increased political commitment and an acceleration of technological innovation, global emissions are likely to follow a medium to high pathway: continuing to increase for the next few decades, and then levelling off or decreasing gradually thereafter' (King et al. 2015). The stress, in highlighting the above statement from a recent publication, is on technological innovation. While nuclear represents a fairly mature technology, much needs to be done to ensure that it is accessible to large parts of the developing world. Given the large capital investments needed in conventional NPPs, many countries have not progressed beyond the research reactors that have been up and running for decades now.

While the allure of low emissions works in favour of nuclear, the important question is to see whether this is the most efficient in terms of the cost of abatement and whether the levelised cost of electricity works in favour of nuclear power generation.

Two studies provide a view on this aspect of SMRs. Iyer et al. (2014) suggest that a scenario with SMRs results in lower costs than a scenario without SMRs. When both large NPPs (conventional) and SMRs compete for the market, SMRs are beneficial only under assumptions of significant advancements and reduction in costs. In another study, Carless et al. (2016) carry out a life cycle analysis of GHG emissions associated with SMR and conventional NPPs. If the nuclear fuel cycle is excluded, both have similar levels of GHG emissions. However, if the nuclear fuel cycle is included, then large reactors (like the AP1000 considered in the study) have 9% lower emissions than SMRs.

Apart from cost and emissions, there are others benefits that an SMR could provide over conventional nuclear plants. One of the features highlighted by proponents of the technology is the ability to provide process heat and use surplus heat for local desalination purposes. Local water treatment in rural areas/clusters is not prevalent today, and there is a need for improved provisions of water. Conventional energy, by way of biomass or LPG is predominantly used by households to heat water and purify it. This can certainly offset a lot of the fossil fuel use in those pockets where desalination and heating requirements are prevalent.

Some manufacturers (such as NuScale) also suggest that SMRs could work hand-in-hand with RE (large grid connected ones). As mentioned earlier, integration of variable RE into the grid is a significant challenge. The requirements of deep-decarbonisation across the globe imply that the penetration of RE will have to increase. In such a scenario, it is necessary to reimagine a system where we think beyond base load, and think of what will provide the balancing during variability. As per the manufacturer's specifications, the multimodule nature of the plant, the manoeuvreability, and '*steam turbine bypass allows the plant output to be varied over days, hours or seconds*' (NuScale 2016). Natural gas power plants have the ability to be load following but at a significant cost—both financial and carbon. The efficiency of fossil fuel-based generation stations reduces significantly when cycled to lower load levels and increases the carbon emissions associated with each unit of electricity produced.

5.2.4 Cost of Nuclear Power, Financing and the Allure of SMRs

Several studies that have analysed the historical trends of the costs of nuclear energy have shown that there has been an escalation in costs for conventional NPPs. 'This lends credence to the arguments that suggest that nuclear energy will never be competitive given that learning and scale has not translated to reduced costs' (Lovering et al. 2016; Hultman et al. 2007; Cooper 2014). 'A significant share of the cost of nuclear power is on account of delays and cost overruns which could range from 15 to 54% of the overall cost' (Lovering et al. 2016). Additionally, it must be pointed out that a significant portion of the increase in costs overtime are also on account of stricter regulatory requirements over time, which has not necessarily been imposed on other generation technologies such as coal-based thermal sources. In some sense, the externalities of coal-based generation are borne by the public at large while those of nuclear are paid for, and there is a regulatory allocation to abate any externalities (a fund for decommissioning, waste management and liability cover).

SMRs represent a viable option to decrease the average capital at risk in the nuclear business. As argued in Barkatullah (2011), reduced plant size and complexity and design simplifications should allow for leaner project management, better management of supply chain and keep a tighter control on overall project costs. With very few SMR projects under construction, there is little data to suggest that economies of scale and reductions associated with co-siting (in existing facilities) will actually materialise. A study in Italy by Politecnico di Milano concludes that the though SMRs do not exhibit the classic economies of scale benefits of NPPs, there are other benefits associated (e.g. modular installations and lower upfront capital risk) with SMRs that could enable their competitiveness. The 'cash flow profile for SMRs is smoother (on account of sequential commissioning of units) and upfront investment is lower so could be more attractive for investors' (Vujic et al. 2012; Boarin et al. 2010). While such studies only provide a theoretical estimate of what the costs are likely to be, these are early indicators that the technology does hold promise and can address some of the challenges of large capital-intensive conventional LWRs that are being commissioned across the globe.

5.2.5 The Physical Risk

My own interest in SMRs started during a 2011 study I was involved in, to look at the feasibility of metropolitan siting of nuclear reactors and specifically the case of Singapore. Other than long-term contracts to source gas from Indonesia and Malaysia, there was seemingly no other way to increase energy security for the import-dependent island state. Renewable energy was not an alternative to a land starved economy, where reclaiming land from the sea has been the strategy to accommodate the constant needs of industry, housing and other strategic assets. Nuclear energy was considered an option worth exploring. In 2010, the Prime Minister (of Singapore) recognised that nuclear energy, while posing additional challenges to Singapore (on account of the small size of the country), cannot be dismissed altogether. He was of the view that 'the country must keep pace with new developments in the technology' (Loong 2010).

Many argued that it was 'impossible to find the necessary sites to locate a handful of GW class reactors of the standard design in such a small city state like Singapore (with a 27 km × 50 km expanse at the farthest point) and that even the basic exclusion zone criteria for conventional reactor units could not be met' (Palmer et al. 2010). In order to satisfy these requirements, it is sometimes thought that nuclear power plants have to be in locations far removed from centres of population (Ganesan 2012). While no plants have been built in city centres, for various reasons, there are many power stations located close to cities. For example, the Barsebäck nuclear power station in Sweden is close to large urban agglomerations. It is 20 km from downtown Copenhagen (in Denmark) and 30 km from Malmö (in Sweden). While neither of the units at the station is in operation today, they were operational for over two decades. The decision to shut these was political, rather than environmental or technical (Palmer et al. 2010). A survey of selected reactors across the world and the population distributions around them provides a more convincing proof to the claim that reactor sites are not (always) that removed from human settlements. In East Asian economies that rely heavily on nuclear, there are multiple reactors that have more than 3 million people living within a 30 km radius of the NPPs (Butler 2011). Contrary to what one would expect, settlements have sprung around established reactors because of the spillover effects that an energy installation brings to the local area spurring businesses and industries (Ganesan 2012; Aldrich 2011). If SMRs can provide sustained energy to local communities, it is very likely that they too could have this effect of coalescing population around the facility, especially in energy deficit economies like India.

This discussion on metropolitan siting is a useful analogy to understand the potential acceptance of SMRs in smaller agglomerations. Living in the vicinity of a plant—even a large one—is clearly not a barrier to the entry of SMRs whose compactness and small size is one of the key attributes highlighted by proponents. Conventional reactors will, for want of public approval and a sense of health risk (whether perceived or real) associated with them, not find acceptance in densely populated areas. Many Asian countries are increasingly finding it difficult to site conventional reactors in sparsely populated areas, particularly in the aftermath of the Fukushima disaster. It will be prudent to ask the question—Will SMRs be able to stem the backlash? On what grounds, will new nuclear technologies (like SMR, FBRs, fusion?) hold merit?

What are the options to address the challenges brought out by a large number of reactors, which might be installed in proximity to population centres? Many designs propose construction to be entirely underground as this automatically brings down the risk of radioactive release. The LWR-based designs have reduced demands for a containment structure. However, 'the design, which has many components integrated into the reactor vessel, could also make maintenance more difficult should one of them fail' (Vujic et al. 2012). In the gas-cooled reactor designs (of SMRs), passive safety provisions are limited as the system does not have an active coolant outside the ceramic coating of the fuel assembly to take away heat at temperatures higher than 1600 °C. Another possible limitation is also the reaction between graphite and oxygen in the event of entry of air. Liquid metal cooled reactors such as those with a lead alloy as coolant would be installed in a tank of water that would draw away excess heat and provide shielding. These are 'some of the features highlighted in literature that serve to reinforce the view that SMR designs are inherently safer but there are many questions still open on the cost that such design features impose and the efficiency of use of fuel in such designs' (Sovacool and Ramana 2014).

5.3 Risk Perception and Technology Maturity

The first coal plants that came to be (in the mid-1880s) had an efficiency of only 1.6%. This then improved to 15% in a relatively short period, but thereon further improvements came slowly. It was not until 1940s that flue gas cleaning for particulate matter was introduced. Flue gas desulphurisation and catalytic scrubbing (for NOx removal) came up only in the 1970s and 1980s. In the developed world, where the value for life (adjusted for quality and morbidity) is much higher than in the developing world, a bulk of the existing power plants came online before 1980 (Termuehlen and Emsperger 2003). It would be a worthwhile exercise—exploring the historic morbidity and mortality related costs that coal-fired power plants imposed on society. Despite the huge burden of disease, coal-fired power plants thrived and were the drivers of industrial growth across much of the world. While technologies exist today to abate these pollutants, the persistence was due to the long leash provided to the technology. Admiral Hyman Rickover, the father of the modern reactor program, imagined that a large nuclear fleet consisting of thousands of reactors would be powering the world economy, but this dream, along with the accelerated development of nuclear technologies, came to an abrupt end with the Three Mile Island and Chernobyl accidents. Risks outweighed the benefits of the technology.

Academic reviews of various designs of SMRs have suggested that the new small reactors have no specific technical and regulatory barriers that cannot be overcome, and these are unlikely to be the reasons that will preclude their growth and acceptance (Vujic et al. 2012). The delivered cost of energy is one of the important factors that determine public acceptability. In theory, SMRs could be cost competitive with large LWRs, the classic chicken and egg problem exists with SMRs. The USP of SMRs is the modularity, and the cost reductions are envisaged only when the production of these modules is in large numbers. Given the regulatory delays and slow pace of build-up of interest in the technology, how will a large demand materialise that will then result in a price drop, which is the precursor to a large demand coming up in the first place.

Environmental and health effects of nuclear power depend on a scientific measure of risk, whereas public fear derives from perceived risks. As articulated beautifully by William Clark, the risk that society attributes to nuclear energy is not just the uncertainty associated with outcome (violent one at that) or the toxicity of substances that are released in the event of an accident. It is perhaps related to the perceived inability to respond in a satisfactory manner to the various complexities associated with the real world (Schwing and Albers 1980). Coping with such notions of risk amounts boils down to a management problem and ultimately relies on flexible institutions and individuals and how free are we to experiment with alternative forms of coping. As a start, clarity that nuclear energy presents a viable option is necessary, before we dismiss it outright.

According to a survey by the IAEA (2005), people appreciate the value of nuclear technology. While '62% of those surveyed supported the continued use of existing plants, 59% were not in favour of new plants being constructed' (IAEA 2005). Research from Japan suggests that when people assess the value of nuclear power at a general level, they consider both the risks and benefits that the project would yield; however, when a decision on siting is to be made, the perceived risks become far outweigh the potential benefits that the project would bring, to the local community (Macintosh 2007). These represent some sort of a divergence in the minds of the public when it comes to crucial questions on the future of nuclear. Public opinion on nuclear energy is dependent on the context on which the questions are asked, it vacillates and is easily influenced and as a result represents a trade-off between needs and perceived risks (Bulleting of the Atomic Scientists 2016). It is surprising to note that in a few of the surveys, a large share of the respondents said that nuclear has a large GHG footprint and even those who were aware of the low GHG footprint did not know that it contributed to the largest share of generation from clean-energy sources.

The progress that we have witnessed over the last two centuries following the industrial revolution is nothing short of phenomenal. From a simple observation by James Watt, who harnessed the motive power of pressurised steam, to the path-breaking discovery of nuclear fission, we have come a long way to harnessing the most compact source of power that is available on the planet. While SMRs are still in their concept stage and the efficacy is yet to be proven, it is only fair that we give new technologies their due importance and time to reach maturity—improve safety, performance and cost-effectiveness. Until such a time, developing Asia (China, India) and developed Asia (Korea and Japan), which have built-up capacity to manage a civilian nuclear program, must remain invested in the technology so as to reap benefits as and when it becomes economically feasible.

The need for technology leadership and maintaining hegemony over the nuclear domain is the key to driving the nuclear agenda in the west. An apt example to cite here is from the frenzy days of the Cold War. At the peak of the space race, the US government was spending up to 4.5% of the federal budget on the space program in a bid to outdo the Soviets who had outdone them by sending the first satellite and human being into orbit. The race to put man on the moon (the MoonShot program that Kennedy launched) saw 12 men in a span of three years (from the first moon-landing) set foot on the moon. In the 43 years since, not a single manned mission has gone past the low-earth orbit (LEO). It is not clear what led to this decline in enthusiasm. However, with the advent of private capital into the space industry, the interest in human space travel is once again gaining momentum. Space is certainly the final frontier—unregulated and free for all. Nuclear, unfortunately, is firmly rooted on earth. Regulation and delays associated with it have virtually sealed the fate of the conventional nuclear industry. Perhaps a breakthrough in experiments such as breeder reactors, fusion or even a successful SMR deployment could resurrect the fortunes of the sector. It is fitting though, that Voyager 1, which is now the farther man-made object from Earth, is still being powered by a nuclear-powered source (radioisotope thermoelectric generator)—nearly 40 years after being launched. Small modular reactors certainly have a role to play in their niche areas. They may not replace large reactors or other power sources anytime soon, but due consideration for the basic tenets promoted by the technology must be given—increased safety, free from regular maintenance and ideal to serve communities with large demands (of heat and power), and located remotely enough to not warrant the extension of the grid.

5.4 Conclusions

For the SMR concept to succeed in Asia, it must satisfy the primary requirements of increased safety (as compared to conventional NPPs), simplicity in construction, operation and must work out being comparable in economic terms with conventional NPPs (after factoring in the pros and cons of economies and financing). SMRs could potentially be taken up by small utilities or electricity cooperatives which cater to the needs of a small community, but one which is too large to rely only on RE. Remote locations, where maintenance-free solutions are required could also be served by SMRs. The limited evaluation of the technical capabilities of SMRs suggest they could offer increased safety by incorporating various aspects of safety that are commonly invoked, such as passive heat removal and the ability to fabricate under tight supervision that will result in inherently safer units being installed.

Siting restrictions are less for SMRs, and as shown earlier in the chapter, millions of people live in the vicinity of large reactors and with no significant opposition or limitation to the quality of their lives. SMRs have a lower land requirement and provide co-siting advantages, and does away with long delays in land acquisition and rehabilitation and resettlement (R&R) of the impacted population. More importantly, the financial view of SMRs is more positive on account of lower capital requirement and the ability to keep construction costs on a tight leash. However, the long run costs (as measured by an LCOE) associated with SMRs will be clear only when multiple installations and significant manufacturing capacity has been established. SMRs certainly have the potential to address the issue of energy access and provide for productive use of energy in remote communities and can also address some of the issues of variability that are associated with renewable energy installations. Energy security is not necessarily addressed by SMRs (when compared to RE or conventional NPPs) with increased reliance on imported modules, that are unlikely to be manufactured domestically for want of economies of scale and specialised manufacturing and materials. On risk perceptions and public acceptance of new technologies, one can conclude that familiarity and awareness are crucial to establishing objective support (or opposition) and for a correct weighting of benefits and costs of a technology.

There are bound to be technical and institutional challenges that will arise as and when SMRs come up for approval and licensing in various countries. Most significantly, the availability of a trained human resource base, that can help manage the installations and ensure smooth functioning, is one aspect that Asia needs to give importance focus to—especially those countries that do not have an active civilian nuclear energy program. Educating the population of the pros and cons and ensuring communities are aware of the opportunities (and risks) that open up as a result of improved energy availability is an important first step in considering new and disruptive technologies such as SMRs.

References

ADB. (2014). Energy security, sustainability and affordability in Asia and the Pacific. Available at: https://www.adb.org/sites/default/files/publication/42782/ewp-401.pdf.

Aldrich, D. (2011). The future of nuclear energy in Japan (Interview). August 1, 2011.

American Nuclear Society. (2010). Interim Report—President's Special Committee on SMR Licensing Issues.

Barkatullah, N. (2011). *Financing nuclear power projects: Challenges and IAEA assistance in capacity building*. Vienna: IAEA.

Bery, S. et al. (2016). Energizing India—towards a resilient and equitable energy system. Sage.

Boarin, S., Mancini, M., Ricotti, M., & Locatelli, G. (2010). *Economics and financing of small modular reactors*. Milan: Politecnio di Milano.

Bulleting of the Atomic Scientists. (2016). Bulletin of the atomic scientists. Available at: http://thebulletin.org/public-opinion-nuclear-energy-what-influences-it9379. Accessed March 26, 2017.

Butler, D. (2011). Reactors, residents and risk. *Nature*, April.

Carless, T. S., Griffin, W. M., & Fischbeck, P. S. (2016). The environmental competitiveness of small modular reactors: A life cycle study. *Energy, 114,* 84–89.

Census. (2011). Census of India. Available at: http://www.censusindia.gov.in/2011-Common/CensusData2011.html. Accessed March 20, 2017.

Chu, S. (2010). Wall Street Journal. Available at: http://www.wsj.com/articles/SB10001424052748704231304575092130239999278. Accessed September 6, 2016.

Cooper, M. (2014). The economic failure of nuclear power and the development of a low carbon electricity future: Why small modular reactors are part of the problem and not the solution.

Ganesan, K. (2012). Civilian nuclear power in Singapore—Technical analysis of siting options (Unpublished master's thesis). Singapore: National University of Singapore.

Hultman, N. E., Koomey, J. G., & Kammen, D. M. (2007). What history can teach us about the future costs of US nuclear power. *Environmental Science and Technology*, (April), 2088–2093.

IAEA. (2005). Global nuclear survey: Public support for new power plants remains tentative. Available at: http://www.iaea.org/Publications/Reports/gponi_report2005.pdf. Accessed January 3, 2012.
IAEA. (2014). Advances in small modular reactor technology developments. Available at: https://www.iaea.org/NuclearPower/Downloadable/SMR/files/IAEA_SMR_Booklet_2014.pdf.
IAEA. (2015). *Climate change and nuclear power*. Vienna: International Atomic Energy Agency.
IEA. (2014). *Energy technology perspectives—harnessing electricity's potential*. Paris: International Energy Agency.
IEA. (2015). *World energy outlook*. Paris: International Energy Agency.
Ingersoll. D. T. (2015). Small modular reactors—Power fad or future? Woodhead Publishing.
IPCC. (2014). Fifth assessment report—Summary for policy makers. Available at: https://www.ipcc.ch/pdf/assessment-report/ar5/wg3/ipcc_wg3_ar5_summary-for-policymakers.pdf. Accessed August 20, 2016.
Iyer, G., Hultman, N., Fetter, S., & Kim, S. H. (2014). Implications of small modular reactors for climate change mitigation. *Energy Economics, 45,* 144–154.
Jain, A., et al. (2015). *Access to clean cooking energy and electricity—Survey of states*. New Delhi: Council on Energy, Environment and Water.
King, D., et al. (2015). *Climate change—A risk assessment*. Mumbai: Foreign Commonwealth Office, UK.
Kuldeep, N., et al. (2016). *Energy storage in India: Applications in the renewable energy segment*. New Delhi: Council on Energy, Environment and Water.
Loong, L. H. (2010). Singapore international energy week. Available at: https://www.siew.sg/docs/default-source/default-document-library/address-by-prime-minister-lee-hsien-loong.pdf?sfvrsn=0. Accessed November 10, 2016.
Lovering, J. R., Yip, A., & Nordhaus, T. (2016). Historical construction costs of global nuclear power reactors. Energy Policy, 371–382.
Macintosh, A. (2007). Siting nuclear power plants in Australia—Where would they go? The Australia Institute.
NEI. (2016). TVA files historic application for small modular reactors. Available at: http://www.nei.org/News-Media/News/News-Archives/TVA-Files-Historic-Application-for-Small-Modular-R.
NuScale. (2016). Why SMR—Diverse applications. Available at: http://www.nuscalepower.com/why-smr/diverse-applications. Accessed August 20, 2016.
Palmer, A. et.al. (2010). Nuclear power in Singapore. IES Journal Part A: Civil & Structural Engineering.
Pannier, C. P., & Skoda, R. (2014). Comparison of small modular reactor and large nuclear reactor fuel cost. *Energy and Power Engineering, 6,* 82–94.
POWERGRID. (2012). Green energy corridor report release. Available at: http://www.powergridindia.com/_layouts/PowerGrid/User/PressRelease.aspx?PRId=2&LangID=&PId=277. Accessed September 20, 2016.
PWC. (2013). Power in Indonesia. Available at: http://www.pwc.com/id/en/publications/assets/electricity-guide-2013.pdf.
Schwing, R., & Albers, W. (1980). *Societal risk assessment: How safe is safe enough?*. New York: Plenum Press.
Sovacool, B., & Ramana, M. (2014). Back to the future: Small modular reactors, nuclear fantasies, and symbolic convergence. Science, Technology and Human Values, pp. 96–125.
Stanculescu, A. (2013). RELAP5-3D quarterly newsletter. Available at: http://www4vip.inl.gov/relap5/news/2013/q2/2013-2q-r5-3d-smr-version4.pdf.
Termuehlen, H., & Emsperger, W. (2003). *Clean and efficient coal-fired power plants: Development toward advanced technologies*. s.l: American Society of Mechanical Engineers.
Vujic, J., Bergmann, R. M., Skoda, R. & Miletic, M. (2012). Small modular reactors: Simpler, safer, cheaper? Energy, 288–295.

Walton, D. (2016). Renewable energy on Islands. Available at: http://www.leonardo-energy.org/blog/renewable-energy-islands.
WNA. (2016). Asia's nuclear energy growth. Available at: http://www.world-nuclear.org/information-library/country-profiles/others/asias-nuclear-energy-growth.aspx.

Author Biography

Mr. Karthik Ganesan is a Research Fellow at CEEW. His research focus includes the development of long-term energy scenarios for India and energy efficiency improvements in the industrial sector in India. He also leads an independent effort to assess GHG emissions from the industrial sector and improving transparency in reporting. In addition, he supports on-going work in the areas of power sector performance and reform and energy access indicators for rural Indian households. He also carried out a first-of-a-kind evaluation of the impact of industrial policies on the RE sector in India.

His published works include Factors Affecting household satisfaction with electricity supply in India (Nature), Energising India—Towards a Resilient and Equitable Energy System (SAGE, book), Rethink India's Energy Strategy (Nature), Co-location opportunities for RE and agriculture in North-western India: Trade-offs and Synergies (Applied Energy), Valuation of health impact of air pollution from thermal power plants (ADB), Technical feasibility of metropolitan siting of nuclear power plants (NUS), Prospects for Carbon Capture and Storage in SE Asia (ADB).

Karthik has a Master in Public Policy from the Lee Kuan Yew School of Public Policy at NUS, Singapore. His prior educational training resulted in an M.Tech in Infrastructure Engineering and a B.Tech in Civil Engineering from IIT Madras (Mr. Karthik can be contacted at: karthik.ganesan@ceew.in).

Part II
Public Perception

Chapter 6
The Clash of Risk Perceptions: Reconciling "the Publics" and "the Experts"

Catherine Mei Ling Wong

6.1 Introduction

Advances in scientific knowledge have never been a guarantee for societal acceptance of new technologies or scientific discoveries. From biotechnology to renewable energy and nuclear power, public resistance to the development and/or application of these technologies have repeatedly challenged the assumption that more knowledge and more advanced technologies translate into less risk and, thereby, greater public acceptance. What we have seen, in fact, is the opposite effect of greater unknowns, more uncertainty, and intensified moral dilemmas around the distribution of risks and benefits.

This paradox between advances in science and technology and growing scepticism in society is most visible in the polarisation of risk perceptions between "experts" and "the public". It is this relationship that has dogged the development and operation of nuclear power plants across the world since its inception. In spite of the growing body of research into the social psychology of public risk perceptions, these works have done little to alleviate the rift between the nuclear industry and the public across the world. This calls for a different perspective; a deeper understanding of where risk perceptions come from; and a broader approach to understanding the apparent clash of risk perceptions between "experts" and "the publics".

Indeed, such categorisations as "the public" and "the expert" are highly problematic. An "expert" is an employee or scientist of a larger organisation, but also often simultaneously a parent, a daughter/son, a friend, etc. Similarly, a lay farmer is simultaneously an expert in agriculture, soil ecology and biodiversity, a parent, a daughter/son, a friend, etc. It is, therefore, not unusual to find scientists among the

C.M.L. Wong (✉)
University of Luxembourg, Esch-sur-Alzette, Luxembourg
e-mail: catherine.wongml@gmail.com; catherine.wong@uni.lu

N. Janardhanan et al. (eds.), *Resurgence of Nuclear Power*,
https://doi.org/10.1007/978-981-10-5029-9_6

public opposing nuclear energy on the one hand, and laypersons supporting nuclear energy on the other, depending on what values they perceive to be under threat.

This chapter unpacks these problematic categories in order to better understand where key stakeholders on both ends of the nuclear debate stand on key issues and why. By presenting original data collected from interviews with senior executives of the nuclear industry and civil society stakeholders in India, the chapter presents two key findings. Firstly, it uncovers the underlying processes that account for public perceptions of nuclear risks that are less about facts than it is about feelings and memories. Secondly, it reveals specific points of convergences and divergences between and within the nuclear establishment and Indian public on fundamental issues. In doing so, the goal here is not simply to present "the other side of the story" and thus to emphasise polarisations between "the expert" and "the public". Instead, the intention is to explore the full range of alternative perceptions evident through the interviews and thence to unpack and challenge practices through which nuclear risk comes to be seen as a straightforward clash of disparate risk perceptions between two homogenous groups of people.

6.2 A Note on Methods and Data Collection

The analysis in this chapter draws on 32 in-depth semi-structured interviews that were conducted in 2012 with the senior executives of nuclear power corporations (both Indian and International corporations engaged in joint ventures with India); government agencies in charge of implementing and regulating nuclear energy, i.e. the Atomic Energy Commission (AEC), Department of Atomic Energy (DAE), the Atomic Energy Regulatory Board (AERB) and the Ministry of Environment and Forestry (MoEF); civil society representatives and senior scientists from other research institutions who have worked with the nuclear establishment. For the purpose of this chapter, participants have been categorised as follows:

"nuclear participants": these are nuclear scientists who (currently and formerly) occupied senior executive positions in the nuclear organisations listed above. Their areas of expertise ranged from nuclear fuel cycle, metallurgy, nuclear engineering, etc., to chemical engineering, finance and economics.

"non-nuclear participants": these are civil society representatives that included scientists from other disciplines like ecology, biology, law, sociology, political science, as well as nuclear physics. This category also included community-based organisation representatives, lawyers and journalists. Participants in this category were critical of nuclear power but not anti-nuclear per se, and are, therefore, distinguished from the next category of participants.

"anti-nuclear participants": these are civil society representatives that include activists, journalists, doctors and freelance researchers who have done research on communities living next to nuclear facilities in India. Participants in this category were strictly anti-nuclear in their position.

In order to keep discussions anchored to the central investigative theme of the research, all participants were asked a core set of questions about their personal journey to, or interactions with, the nuclear industry and what they thought were the risks of nuclear energy. Questions were kept general and open-ended, allowing space for participants to draw their own associations to other events, objects, memories or reflections. This led to some unexpected findings presented in the following sections.

Interviews were analysed using the inductive method of thematic analysis, which is the search for emergent themes that are important to the description of the phenomenon under study (Fereday 2006: 82). The QSR NVivo software was used to manually code the interview transcripts in order to identify emergent dominant themes as well as marginal themes which are less visible. This is when only one or a few participants mentioned an idea, detail or association that was missed by a majority of participants, but which carries great insight, innovation or significance (see Saldana 2009; Namey et al. 2008; Spradley 1979). The analysis presented here draws only on parts of the data that relate specifically to "public" and "expert" risk perceptions and, in particular, the voices of moderation on both sides of the debate.

6.3 Conceptualising Risk Perceptions

Before entering into a discussion on the empirical findings, a conceptual overview is first necessary. Given the multifaceted nature of risk, research on the way it is perceived, constructed and understood has been eclectic in disciplinary orientation.

6.3.1 Social Psychology and the Problem with Cognition

Some of the earliest works on risk perception come from social psychology which used psychometric methods to quantify risk perceptions. This is done by getting different social groups, primarily (and simplistically) distinguished between technical experts and lay publics to rate a range of dangerous activities or events. Factors like memorability, imaginability, judgment biases and cognitive limitations are deemed to be key variables that influence public risk perceptions, typically leading to conclusions that better information, communication and education of the public would lead to more objective evaluations of risk (see for example Slovic 1987; Slovic et al. 1979).

Such claims have been empirically challenged, not least by the persistence and growing polarisation of risk perceptions between "experts" and "the public" in spite of greater access to information. Theoretically too, the psychology of risk perception has been criticised for its lack of critical reflection on cognition, which is largely assumed to exist a priori in psychometric studies. This negates the ways in which cognition, expert or otherwise, is socially produced. Indeed, numerous studies have shown that decisions about what risks are worth taking, how much risk is acceptable, who should decide and who should be responsible, are socially, culturally and politically mediated (Douglas and Wildavsky 1982; Jasanoff 1994; Beck 2010; Short 1984). Social consent implicit in the prevailing social principles, power structures and internalised value-judgments of one's community, organisation, disciplinary tradition, etc., shape risk perceptions of the individual, whether one is an expert or lay person. Therefore, cognition must be understood within its larger context of social institutions/organisations, the thought collectives embedded within them, and the social order that its members internalise. In other words, cognition cannot be used as an explainer for risk perception, but must be, itself, explained. This leads us to the question of organisations and their role in shaping risk perceptions, particularly of the "experts" that operate within them.

6.3.2 Organisations, Risk and the Problem with Trust

Organisations influence almost every aspect of society, from technology, markets and politics, to social class, religion, identity and social psychology (see Perrow 1991; Short and Clarke 1992). Indeed, the prerogative of organisations to set the parameters and terms of debate for risk decisions on behalf of society was largely left unquestioned until the 1980s when advanced technological complexes and their associated organisations began to unravel. From the Challenger space shuttle disaster and the Chernobyl nuclear disaster in 1986 to the Exxon Valdez oil spill in 1989 and Bhopal chemical disaster in 1984, organisations began to come under increasing scrutiny as a source of risk rather than a bastion of risk deterrence.

Research into organisations found that so-called objective decision-making was, in fact, a highly subjective outcome of social selection, conceptualisation and interpretation (see Reiss 1992; Vaughan 1996, 1999). Organisations were not orderly and rational entities but, instead, messy inter- and intra-organisational relationships where experts disagreed on scientific methods and interpretations of facts; where rationalisations were economically and politically motivated; and where personal experiences and "rule of thumb" defied the neat, technically sophisticated and ideologically neutral portrayal of risk analysis (Short and Clarke 1992: 8–9). Experts, therefore, are equally subjected to the social processes of selective knowledge production, and the evaluation of risk is often not the outcome of objective risk-benefit assessment but subjective political trade-offs.

Indeed, the unravelling of organisations in the post-industrial world led Beck (1992) to postulate that we have now entered into a new Risk Society, in which key institutions of modernity such as science, business and politics are no longer viewed as trustees but suspects of risk making. While they used to be considered trusted managers of risk, they are, today, seen as the sources of risk. This erosion of trust in organisations and the widespread implications for public risk perception have been well documented in the nuclear industry. Freudenburg's (1988, 1993, 2003) survey of public attitudes towards a proposed low-level nuclear waste facility in the USA showed that socio-demographic and value-based ideological factors were, in fact, weak indicators of public risk perception. Far more influential were the degree of trust in (1) science and technology, (2) business capability and (3) the federal government (Freudenburg 2003: 106). These variables more than tripled the levels of concern registered by respondents from 29 to 92% (Freudenburg 1993: 919–924). The failure of experts or specialised organisations to carry out their duty of care and/or their misuse of authority had far greater influence on public risk perceptions and, in fact, had amplifying effects on "actual" risk. Similarly, a later study by Flynn (2003) on nuclear stigma in the USA found that a history of false reporting, scientific manipulation, secrecy, public denials of leaks and incidents, and unethical testing on the environment and human subjects all culminated in present-day public distrust of the nuclear establishment.

Such observations are not limited to the USA alone. Similar conclusions were drawn in the UK by Wynne (1992) in his study of Cumbrian sheep farmers in Sellafield after the Chernobyl disaster. The poor public uptake of scientific knowledge, in this case, was attributed to the clash of intellectual frames between scientists and farmers and the organisational body language of scientific institutions that had, for years, denigrated farmers and the specialist knowledge they possessed of their livestock and local environment. In particular, Wynne (1992: 287) found that the exaggerated sense of certainty and arrogance was a major factor undermining the scientists' credibility with the farmers whose cultural ethos routinely accepted uncertainty and the need for flexible adaptation rather than prediction and control. As a result, in the aftermath of the Chernobyl nuclear fallout, farmers disregarded scientific claims about safe radiation levels and the exposure of their livestock. The lack of trust, therefore, had the detrimental effect of falsifying all facts, regardless of their value or intent. As Law (1994: 170) aptly describes:

> Who will believe what they say when they claim that it is all safe? If you already distrust them, then is not their very appearance of honesty artifice?

Indeed, trust is not simply a matter of goodwill or perception. It has very concrete and material consequences in the ways people respond in a risk event, who they choose to take advice from, what sources of information they choose to believe, and their willingness to change behaviour to reduce risks. The misplaced trust in spurious scientific claims of fringe scientists about climate change and the distrust in evidence-based IPCC reports, for example, show how trust impedes important social, political and technological transformations needed to reduce risks. Misplaced trust can also lead to rippling effects that impact other previously unrelated industries, sectors and policy domains when people, investors, consumers, etc., panic (see Kasperson et al. 2010). Last but not least, the lack of trust leaves development projects vulnerable to exploitation by those seeking to make quick political gains by politicising particular scientific, technological or economic activities. Resultant public resistance incurs significant costs from delayed policy or project implementation, the need for added security personnel and drawn out legal battles, as has been the case in India.

Trust, unfortunately, is often either assumed to exist or misunderstood to be simply the product of more communication, greater transmission of information and better education of the public. While these are important aspects of cultivating a better understanding of risks, one-way communication without meaningful dialogue serves only to deepen distrust of expert claims and further polarise "expert" and "public" risk perceptions. Indeed, public consultations tend to be located at the final stages of decision-making where the primary function of public involvement is to evaluate options that have already been decided by an expert-led process (Healy 2004). This does little, if at all, to build trust among the public or lend legitimacy to the decision-making process.

6.3.3 Risk Governance and Trust-Building

The limitations of conventional forms of public consultation call for new approaches to public involvement in risk assessment and risk management. A major challenge, however, is in how best to integrate the public in decision-making given that different individuals/groups have different and often conflicting interests?

A governance approach attempts to address this by reframing the logics of public participation: if risk management is understood as an agreement on priorities and trade-offs that are considered fair and acceptable (and not just a calculus of probabilities), then a wider set of stakeholders, values and knowledge pools must be enlisted. Certainly, it is not practical for all members of the public to be involved in every risk decision, but there are a number of ways to determine when and how much public participation should be incorporated. The Integrative Risk Governance Framework developed by Renn and colleagues offers one approach using a "stakeholder involvement escalator" (Klinke and Renn 2014; Renn 2014, 2015), which distinguishes among four types of risk problems: (1) linear, (2) complex, (3) highly uncertain and (4) highly ambiguous. The type of risk problem indicates the level of stakeholder involvement needed ranging from largely expert involvement in dealing with linear risks to extensive involvement of technical and social scientists, affected public stakeholders and civil society in highly complex and ambiguous risk problems (see also Renn 2006; Renn and Klinke 2012).

Following the same logics of public participation, but more focused on conflict resolution at a local scale of project planning and implementation, Wong (2015a) proposed a Hybrid Risk Governance Framework composed of five stages: co-selection, co-design, co-definition, co-planning and co-regulation. A selection criteria for participation guides the process from the first stage, followed by a set of deliberative questions to steer discussions in each of the subsequent four stages, with an avenue to return to earlier stages of deliberation when disagreement or conflict arises to re-evaluate and re-negotiate the set of interests and priorities.

What both these approaches seek to do is offer a set of practical formats for organising public participation in decision-making about risk. Importantly, both approaches see public stakeholders not only as valuable sources of knowledge but also important agents for disseminating the results of the governance process and facilitating outreach (Renn 2015: 8). This lends legitimacy to the decisions made and builds trust in the process, closing the gap between expert and lay risk perceptions.

6.4 The Elusive Indian Public

In seeking to close the gap between "the expert" and "the public", a lingering problem one runs into is who "the public" is as a unit of analysis. Indeed, this broad category is highly problematic in India where class, caste, occupation, religion, politics, culture and ethnicity intersect within this single classification. In the Kudankulam Nuclear Power Plant (KKNPP) controversy, for example, the anti-nuclear public is a mixed bag of a dominant mixed Hindu-Christian Nadar community of farmers and middle-class citizens; a Muslim minority; and largely Roman Catholic Mukkuvar fishing community. Each of these groups supported and/or opposed the nuclear power plant at different stages of the project's development for very different reasons.

The Nadar community, made up of mostly farmers and middle-class Indians, were not natural allies of the anti-nuclear movement. They were, in fact, initially supportive of the nuclear project. Seeing it as a boon for economic development in the area, some were even willing to sell their land to the Nuclear Power Corporation of India Limited (NPCIL) at prices considerably below market value with the expectation that they would be given preference in the job and sub-contracting opportunities that would come with the nuclear power plant (Abraham 2013). This initial enthusiasm, however, started to wane when they realised that the nuclear facility would draw large volumes of freshwater from the Pechiparai dam that the Nadar farmers depended on for irrigating their crops. The Muslim community in Kudankulam too was not easily convinced to support anti-nuclear activists particularly due to the post-9/11 political climate. Unwilling to engage in activities that might antagonise the State, the Muslims in Kudankulam were, in fact, reluctant to join the anti-nuclear movement.

The Mukkuvar fishing community, on the other hand, was readily supportive of the anti-nuclear resistance for reasons that were entirely unrelated to nuclear power. Their support for the anti-nuclear movement, in fact, stemmed from the geopolitical tussle between India and Sri Lanka that had, over the years, seen Mukkuvar fishermen regularly detained by the Sri Lankan navy for the purposes of prisoner exchange with India (Gupta and Sharma 2008). These experiences have cultivated a deep distrust in the government by the Mukkuvar community and a sense that the State had little interest in securing the safety of Indian fishermen in the region. This distrust, in turn, spilled over into the Kudankulam nuclear power plant project which was deemed as yet another government project that had no interest in the welfare of local communities.

At a national level, there was also a vast difference in economic and political interests in rural India as opposed to its urban and metropolitan cities. When research participants spoke of "the public", there seemed to be a conflation of "the public" in the cities and "the public" in rural host communities. For participants from the nuclear establishment, the term was used to refer both to people in urban centres and rural village folk. But for anti-nuclear participants, the two groups of publics were quite distinct. The urban public was seen as the ones reaping the benefits of electricity supply from nuclear power plants, while the rural public bore all the risks of hosting and living next to the nuclear plants. Furthermore, rural India was poorly connected to the grid and received little (if any) of the electricity produced by the power plants because electricity generated from power plants are fed into the central and regional grids, which then directs supply to metro cities as a priority. Hence, the urban public was perceived as an ambivalent metropolitan audience, largely apathetic to the concerns of rural village folk. At times, student movements tried to galvanise urban populations around nuclear risks. But these movements were sporadic and driven more by interests in biodiversity conservation and/or anti-US sentiments rather than rural poverty and inequality (see, for example, NDTV 2011b; Sovacool and Valentine 2010; Srikant 2010). Urban-based anti-nuclear movements have also focused more on the proliferation of nuclear weapons than on the potential of a nuclear fallout from nuclear power plants (Srikant 2009: 3).

Hence, the public is heterogeneous and difficult to define. Measuring their acumen towards risk and levels of trust based on individual social or cognitive traits is almost guaranteed to produce misconceptions about how people actually think about risk, whose views they share, and where people might, in fact, agree with each other in spite of other differences as the following sections illustrate. For this reason, this chapter uses the term "publics" to emphasise the diversity of views within this single analytical category.

6.5 The Clash of Risk Perceptions: Roots and Origins

The clash of risk perceptions between the Indian publics and the nuclear establishment is not new and requires little illustration. What has been less well understood is the underlying reasons for what has often been presented as irreconcilable rationalities between two groups. The analysis here, in fact, finds that the points of conflict between "the publics" and "the experts" were less about facts than it was about feelings of distrust. And these feelings stem from a range of historical processes that sometimes occurred (and continue to) in arenas that the nuclear establishment had little control over. Indeed, there were instances where the nuclear establishment was, itself, a subject of local power plays and politicking. Within this history of distrust in the making, five sets of processes were identified in the analysis.

6.5.1 Process 1: Played by Politics

The lack of trust in the nuclear establishment can be partially traced to distrust in local politicians and government organisations as a whole, which have been in the making before controversies around nuclear energy even started to emerge. This was best described by nuclear participant 14:

> Government bodies are trusted little by people. The Central Government is finding it difficult to carry the state government along. Likewise, the state government finds it difficult to convince the locals.

The hybrid nature of the nuclear industry as a State-owned enterprise has meant that the failings of the government (central/regional/local) were often associated with the nuclear industry and vice versa. This was reflected in the data: when anti-nuclear participants spoke about cover-ups, the lapses in justice and lack of transparency in the nuclear industry, it was difficult to differentiate whether they were referring to the government or to the nuclear industry. Indeed, they often did not distinguish between the two. In other instances, the failure of the MoEF to conduct credible environmental impact assessments (EIAs) in other industries was used as indications that EIA approvals given to the nuclear industry were also unreliable (see EPW 2012; Jebaraj 2012).

Central–state politics in India have also eroded trust in local politicians who used anti-nuclear sentiments to win votes, only to reinstate nuclear projects after the elections. Tamil Nadu's late Chief Minister, Jayalalithaa, for example, was a vocal supporter of the anti-nuclear protest before the 2011 local elections and halted all work on KKNPP until the Central Government and the nuclear establishment could allay the fears of the local communities living close to the plant (NDTV 2011a). As non-nuclear participant 19 recalled:

> The provincial government here [referring to the Tamil Nadu government], was pro-people actually. So it became a shield for them. So it told the Central government not to use any force to crush the [anti-nuclear] movement.

But several months after Jayalalithaa won the state elections, she reneged her position and allowed construction in KKNPP to resume. When local protest broke out against the commissioning of the first KKNPP unit in September 2012, Jayalalithaa did nothing to prevent the use of force by Indian police and Special Forces on unarmed civilians, including women and children.

Indeed, this was not an isolated case. The farming and fishing communities in Haripur, West Bengal, were similarly played out by the All India Trinamool Congress (AITMC) headed by Mamata Banerjee. Before the 2011 elections, she too was a strong supporter of local protest against the Central Government's plans to build six 1,000 MW Russian reactors in Haripur. Using her political clout, the nuclear project was shelved. This was heralded as one of the rare instances where the voices of the local farming and fishing community triumphed over the large, powerful nuclear establishment (Roy 2010; Times of India 2013). But barely three years after Banerjee's party came to power, talk of the nuclear project resurfaced and announcements were made in the press that nuclear plans for Haripur would resume. Similarly, the six units of 1,650 MW reactors planned for construction in Jaitapur in the western state of Maharashtra have the same trappings of political manipulation by state politicians. As non-nuclear participant 27 noted:

> Their [referring to the political party in Jaitapur] intention is politics, nothing beyond that. They are supporting the agitators…they say that this Jaitapur Plant is not safe…[but] this [Kudankulam] plant is safe, which is nonsense…I mean you oppose Jaitapur, then why are you supporting Kudankulam?

Nuclear power plants, therefore, have been routinely used by state/local politicians as political currency to win votes, only to backtrack on earlier election promises and even facilitate state aggression against civilians. These are often processes beyond the control of the nuclear industry, and because it has been ill-equipped to navigate such vicissitudes of local politics, nuclear power plants have become a proxy for public vitriol and retaliation against the state or regional government.

6.5.2 Process 2: Promises Not Kept

That said, the nuclear industry has also been responsible for public distrust in its institutions. Promises made to local communities for agreeing to host nuclear power plants have, in the past, not been kept. Villagers affected by the Tarapur nuclear plant, for example, were promised generous compensation for their relocation. But according to non-nuclear participant 27, they were only paid "some pittance" and promises of more stable employment opportunities at the plant were also unfounded:

> I mean they don't keep their word…once the plant becomes operational, all these promises will remain on paper…There are no employment opportunities…during construction, they are basically engaged…as a contract worker, and not much regular employment really.

The same story could be found in Gujarat where Adivasi (indigenous) communities hosting the Kakrapar Atomic Power Station (KAPS) nuclear plants were promised jobs, utilities and schools when operation of the plants started in the early 1990s. Some 25 years later, the local school that was demolished for the project had not been replaced and nothing had been done to refurbish the existing village primary school which was derelict. Officials came to test the water from the Kakrapar dam that flows into the Tapi River supplying water to surrounding villages, but no information was communicated to the local communities about safety and level of radiation in their water source. Some villagers even invested in technical education and engineering degrees for their children so they could benefit from the new employment opportunities at the nuclear plant. But not only were jobs for local villagers scarce, but young graduates from the villages found that preference was given to children of employees at the plant (Gaikwad 2014).

6.5.3 Process 3: Bhopal's Ripple Effect

Another event that was not originally connected to the nuclear industry but had deep reverberations in the latter was the 1989 Bhopal disaster. While the accident occurred in an entirely unrelated industry (the Union Carbide factory was producing pesticide), the failure of regulators to adequately monitor the industry and the subsequent failure of the justice system to enforce accountability destroyed public trust in India's regulatory system and large corporations beyond the chemical industry. In the years, indeed, decades that followed the disaster, evidence began to surface that the Supreme Court and Reserve Bank of India had, together, downgraded a number of death cases to injury and illness and suppressed the total number of deaths in order to reduce the total amount of settlements to the victims (Muralidharan 2004).

The ripple effects of the Bhopal tragedy in the nuclear industry were clearly evident in the interviews with non- and anti-nuclear participants, all of whom made references to the Bhopal disaster. In particular, the trends of regulatory failure, lack of organisational duty of care, institutional miscarriage of justice and the marginalisation of the poor in the case of Bhopal were said to be already evident in the nuclear industry. As non-nuclear participant 22 expressed:

> Even like the Bhopal gas tragedy…even there the company has not taken the responsibility of the devastation even after 27 years they have not been properly compensated. And even here [referring to nuclear power], the Government is repeatedly saying "It is safe, it is safe", but the people are not ready to believe it…the Bhopal gas tragedy already set a precedent.

It is, however, important to note that the association with the Bhopal disaster was not entirely lost on the nuclear industry. Indeed, a number of senior nuclear executives interviewed also made references to the event and expressed concern over how to disassociate themselves from it. This opens important discussions into how the industry can do things differently and learn lessons from Bhopal.

6.5.4 Process 4: Undervalued Expert Knowledge

Similar to the case of Cumbrian sheep farmers in the UK (Wynne 1996) and the nuclear industry in the USA (Flynn 2003), the local specialist knowledge rural folk possess has been undervalued and underutilised by (nuclear) scientific institutions in India. This is, in fact, a missed opportunity.

Local specialist knowledge has on a number of occasions proven to be more accurate than expert scientific knowledge in India, especially in cases of ecosystems management (see, for example, Gadgil 2014). Even among the nuclear participants in the study, some admitted that local villagers have expertise in their environment from generations of inherited knowledge and decades of lived experience in the area which nuclear scientists could never compete with. Local farmers, for example, have better knowledge about weather patterns, behavioural patterns of livestock and wildlife in the area, biodiversity and the use of particular waterways and natural features for daily needs, rituals and religious practices. If incorporated into the planning and design of the nuclear facility, local expert knowledge sets can play an important role in improving safety and the technical integrity of a nuclear plant in a manner that also maintains (as best as possible) the integrity of local ecosystems, livelihoods, culture and religious practices.

Indeed, contrary to common assumptions of ignorance, rural communities are, in fact, extremely well informed about regional and international events through local non-English newspapers, which are widely read and discussed at the village level. Research participants who work with rural communities also pointed out that even though village folks in India are mostly illiterate, they do understand scientific reasoning. They also highlight that many rural families have children who are educated and can translate and interpret more complex scientific information for their parents. Furthermore, in states like Tamil Nadu, rural folks are predominantly Christian and particularly well versed with the English language. They also have access to a wide network of NGOs which are connected to an even wider network of scientists within and outside of India, including physicists, biologists and geologists. The assumption that rural folks are uneducated, and thereby, ignorant and unable to grasp basic scientific reasoning is, therefore, a gross misconception.

6.5.5 Process 5: When Targets Create Great Expectations

The nuclear establishment has been known to occupy a special position within India's political infrastructure with direct access to the Prime Minister, a privilege no other scientific institution in India enjoys. But while the alliance between the nuclear establishment and the Government are strong, they do not always share the same priorities, goals and time frames. Politicians were driven by their own political agenda to support prestigious mega projects to boost energy security and economic growth. But this has, on a number of occasions, compelled them to make public announcements of production targets on behalf of the nuclear establishment which even the nuclear industry considers unrealistic. For instance, when asked how the target of making nuclear power account for 25% of the national energy mix was arrived at, nuclear participant 13 responded with an incredulous:

> Well, I'll be very happy if we can get to 25 per cent!

Another example can be found in the mixed signals from State representatives. On one occasion at an ASSOCHAM International Conference on Nuclear Power in 2012, Minister of State, V. Naranyanasamy told journalists that the controversial KKNPP would be commissioned "within 20 days minimum, 30 days maximum". This drew sharp criticism from nuclear participant 11 who said that the minister should not have announced such unrealistic targets to the public which were not only baseless but also put the NPCIL in bad light and affects the nuclear industry's credibility. Indeed, even the Hindustan Times noted that Minister V. Naranyanasamy had said at least 16 times in 18 months that the KKNPP would be commissioned within 15 days (Jayaraman 2013).

That said, it is true that the nuclear industry has been guilty of missing their targets. The first nuclear production target under the 1962 plan projected 20,000 MW in nuclear capacity by 1987. But the actual amount achieved by that year was 512 MW. The DAE significantly reduced the next target for 1990 to 8,000 MW, but the installed capacity that year again fell a long way short at under 1,000 MW. The production target for 2000 was downgraded from 43,500 to 10,000 MW and, even so, only 2,720 MW was achieved (Bidwai 2012: 294). Certainly, there were larger geopolitical processes at work that created political, scientific and technological barriers to the development and production of nuclear power as planned. What is also less known and understood is the distinction between the vision set by the DAE, which tends to be ambitious, and the targets that are eventually sanctioned by the Government, for which much more modest funding and resources tend to be allocated, not least due to competing policy priorities. As one industry insider lamented, even though the DAE's vision of 10,000 MW of nuclear generation capacity by the year 2000 was approved by the Government, annual allocations of State funding to NPCIL from 1990 to 2000 were so poor that reactors under construction could not progress. Likewise, a decade later, the DAE's vision of installing 20,000 MW nuclear capacity by 2020 was sanctioned by the Government but this was, again, not supported by the funding required to meet such a target. This was further compounded by continued import restrictions under the Nuclear Suppliers Group and uranium mining restrictions at the local level.

What these observations illustrate is that while the DAE does enjoy a special position within the State apparatus, it too has to compete for State funding and resources. This disconnection between DAE visions and the targets made possible by State funding reflects a deeper disjunction between the nuclear establishment and the Government's goals, priorities and time frames. But ultimately, regardless of whether visions have been misunderstood as targets, when the DAE announces or the Government approves grandiose production targets, it creates great expectations from the public, whose trust hinges upon the nuclear industry's ability to deliver.

6.6 Divergences and Convergences in Risk Perceptions

The long and complex history of distrust in the making through these five processes, notwithstanding the clash of risk perceptions between "the publics" and "the expert" nuclear industry is neither inevitable nor irreconcilable. Deeper analysis into how participants on both sides of the debate thought about the risks of nuclear power revealed some unexpected divergences and convergences within and between both groups.

6.6.1 Divergences Within

It was found that non- and anti-nuclear participants disagreed with each other on a number of key issues such as the viability of the three-stage (fast breeder) programme, the link between cancer and radiation from nuclear power plants and the root causes for concern in the larger debate about safety (see Table 6.1 for an overview).

Table 6.1 Overview of disagreements among non- and anti-nuclear participants[a]

Issue	Disagreeing participants	Details
Three-stage nuclear programme and root problem of the nuclear industry	Anti-nuclear participant 4 versus Non-nuclear participant 9	Participant 4 called the three-stage programme "basically science fiction" and invariably "unviable". He felt that there was no way nuclear power could be viable and must be scrapped in its entirety. Participant 9, on the other hand, still strongly believed in the promise of the fast breeder and thorium reactors. He considered the root problem of the nuclear industry to be the institutional subjugation of the regulator, the lack of transparency and the import of new foreign reactors, which he saw as unpatriotic and an erosion of India's long and hard-earned intellectual independence since the days of the nuclear embargo.
Causal link between cancer and radiation from nuclear plants	Anti-nuclear Participants 4, 18 and 20 versus Non-nuclear participant 24	Participants 18 and 20 believed that radiation from nuclear power plants was responsible for the high cancer rates in the local community. Participant 4 also believed that the high incidence of Down's syndrome and thyroid abnormalities in some areas in India was linked to radiation from nuclear materials. Participant 24, however, is far more cautious about the links between the high incidence of cancer and radiation from nuclear power plants because uranium mines and nuclear reactors release many other cancer-causing pollutants.
The need for nuclear energy	Non-nuclear participants 9, 17, 19 and 24 versus	Participants 17 and 19 felt that in the context of India's energy resources, nuclear energy could play a role because of India's energy needs. Participant 17 also believed that there were fewer risks

(continued)

Table 6.1 (continued)

Issue	Disagreeing participants	Details
	Anti-nuclear participants 4 and 18; and Non-nuclear participants 22 and 25	involved with nuclear energy relative to other energy-producing technologies. For her, the main risks laid in the way it as managed and not the technology per se. Participant 19 also considered the problem with nuclear energy to be the lack of democratic process in decision-making and the ethical questions around risk and benefit distribution. Participant 9 too felt that despite its risks and weaknesses, nuclear energy was "not something that can be entirely abandoned". Participant 18, however, felt that nuclear technology was inherently unsafe and, therefore, must be abandoned. Participant 4 called nuclear scientists "delusional people" working on an "inherently, unacceptably and irredeemably hazardous technology". There was, therefore, no way to justify the use of nuclear power. Participants 22 and 25 felt that nuclear power had always been dangerous and harmful to human health and that the State should, instead, invest in alternative technologies like solar, wind and biogas.

[a]Not all non- and anti-participants disagreed with each other, and some did not raise any of the issues of interest in this table. Therefore, only the relevant participants are included in this table

6.6.2 Convergences Between

On the other side of the nuclear debate, reflexivity and moderation was a distinct feature in interviews with the nuclear establishment. Indeed, there were instances of convergence between nuclear and non-nuclear participants on issues of transparency, climate change and clean energy.

Some nuclear participants acknowledged that mistakes had been made in the way the industry interacted with the public. They noted, however, that this was, in part, a legacy of the late 1940s when the Atomic Energy Commission was set up under the auspices of the Prime Minister's office and protected by the Official Secrets Act. Reflecting on their institutional history, nuclear participants noted that communication with the public was something the industry never had to consider until 2005 with the passing of the Right to Information (RTI) Act, which somewhat fortuitously coincided with the signing of the 123 Indo-US Agreement. As nuclear participant 7 explained:

> All of us, you know the leaders of atomic energy, we are all from the training school of BARC...we are given training on science and technology of nuclear power...we are not trained in economics...we are not trained on social sciences. So I can name the reactor for you, but I don't know how to convince the guy, you understand?

Nuclear participant 11 also admitted that the nuclear establishment was struggling with transparency today because nuclear power was historically operated on the basis of secrecy:

> Ok, maybe we may need more transparency...as I said, it [the nuclear programme] was combined — the civil and the strategic — everything was combined. So there was a different mindset, a different way of thinking...It's not intentional seriously. I think it never occurred to them.

For this reason, both nuclear participants 7 and 11 emphasised that the industry needed time to make the transition from being a Government organisation working under conditions of complete secrecy for the last 50 years, to a civilian entity today. There is reason to believe that changes are occurring in this respect. Specific positions have been created both at the level of the DAE and at the NPCIL to cater to communication and public outreach occupied by individuals trained in communications and public relations. The efficacy of these changes is a separate debate, but this is at the very least, a step in the right direction. Other research on elite risk perceptions of the nuclear industry in India also found evidence of self-critique and reflexivity. Specifically, Wong (2015b) notes that members of the nuclear establishment acknowledged the need to (1) look beyond producing energy to producing ethics, (2) acquire the social licence to operate and not just environmental and regulatory licences and (3) exercise greater scientific humility and admit mistakes when they have been made and recognise ignorance where uncertainties lie.

Moderation could also be found among the non-nuclear participants. Participant 19, for example, acknowledged that "completely being anti-nuclear or completely being pro-nuclear was not the answer" and putting "sudden brakes" on India's industrialisation process was not desirable either. He felt that "somewhere in between is the answer", but the appropriate measures should be taken by the nuclear establishment and the people given the necessary tools and training to deal with an emergency or disaster. Participant 31 also believed that nuclear energy "might be a necessary evil in the short term...as a means of buying time" until a breakthrough in nuclear fusion technology was made. There was also empathy for the nuclear establishment's position on energy security and global environmental change by participant 17 who felt torn on the issue:

> I'm caught between that [risk from nuclear power] and worrying about climate change, which is as problematic...and could affect people's livelihoods just as badly...and in a funny way, with greater certainty, because it's going to happen. And the more we use coal, the more it's gonna happen.... So, for me right now it is, ok fine nuclear energy within very limited quantities, after you've done due diligence on the sites...Not all the people, but a lot of the community agrees that nuclear energy is one of the options we have to look at, but people are very concerned. We say this with a heavy heart.

Indeed, the role of nuclear energy in mitigating climate change was something that non-nuclear participants generally agreed on. But an important caution was raised by participant 24 who pointed out that climate change is "too big of a problem to be fixed through simple technical fixes". The bigger and more important question to ask, he argued, was "what kind of India" the people wanted and where nuclear technology sat within this larger vision. Without proper reflection on this question, decisions to use nuclear energy to mitigate climate change may result in "the worst of both worlds" where India is left to bear all the risks of using nuclear power and the effects of catastrophic climate change at the same time.

6.7 Conclusion

This chapter has illustrated how risk analysis and risk management in the nuclear industry (or any other hazardous activity for that matter) is never a simple straightforward calculus of probability, consequence and costs. Similarly, risk communication is not a simple matter of "educating" the public and provision of information when the public does not trust the source of information. Ultimately, a project or activity that is considered technologically and economically viable but socially unacceptable will still fail. If the values that society hold dear are either not considered in the decision-making process or considered to be too precious to be traded-off for whatever economic or technological benefit it may bring, technical robustness alone will do little to change public perception and acceptability. The risk perception of "the publics", therefore, should be treated as a representation of the set of values that need to be accounted for in decisions about risk. In practice, however, public participation in decision-making in the nuclear industry has been weak in large part because of simplistic "us versus them" and "pro- versus anti-nuclear" polarisations that compresses diverse views into a single artificial category and locks people into fixed and combative positions.

This chapter challenged these modes of thinking by firstly, illustrating that "the public" is not one homogenous category of people who necessary share the same views. Rather, there are many "publics" that support and/or oppose nuclear power for many different reasons. The underlying reasons for this outwardly simplistic clash could be traced to the five historical processes of distrust in the making that go beyond the nuclear industry itself, implicating the entire network of Central Government and state politics, judicial systems, ministerial functions, scientific institutions and industrial relations. It is in this historical context of distrust that the very appearance of honesty by the nuclear industry has been conceived as artifice and nothing more than an exercise in impression management which breeds more distrust.

This, however, does not imply that conflict and the clash of risk perceptions is inevitable. The crucial lesson to be learnt here is that increasing communication of hard, technical facts to the public will do little to reconcile risk perceptions. Instead, emphasis should be placed on building trust through early and meaningful engagement with public stakeholders, incorporation of social values in decision-making and more equitable distribution of benefits to those most affected by nuclear activities. Indeed, the reconciliation of views and interests is not impossible as the preceding section illustrated. Public views about what exactly is at risk varied extensively; they were not inherently opposed to nuclear power and, in fact, considered it necessary to some extent. Members of the nuclear industry, too, are increasingly cognizant of the need for more public accountability and transparency.

The depolarisation of the nuclear debate, therefore, is possible. But it is more than a matter of discursive reconstruction or manipulation. It requires meaningful deliberation in practice based on the understanding that while it is not always possible to satisfy everyone's goals and aspirations, it is possible for people's views to be swayed by rational arguments and to lay aside particular interests and opinions in deference to overall fairness and the common interest of the collectivity (Lockie and Measham 2012).

References

Abraham, I. (2013). The violence of postcolonial spaces: Kudankulam. In: *Conflict, Security and Development (CSD) Conference: Habitations of Violence*. Hyderabad.

Beck, U. (1992). *Risk society: Towards a new modernity*. London: Sage.

Beck, U. (2010). *World at risk*. Cambridge: Polity Press.

Bidwai, P. (2012). *The politics of climate change and the global crisis: Mortgaging our future*. New Delhi: Orient BlackSwan.

Douglas, M., & Wildavsky, A. (1982). *Risk and culture*. Berkley, Los Angeles, California: University of California Press Berkeley.

EPW. (2012). Cooking up environmental assessments. *Economic and Political Weekly. XLVI, 8*, 7–8.

Fereday, J. (2006). Demonstrating rigor using thematic analysis: A hybrid approach of inductive and deductive coding and theme development. *International Journal of Qualitative Methods, 5*(1), 80–92.

Flynn, J. (2003). Nuclear stigma. In N. Pidgeon, R. E. Kasperson, & P. Slovic (Eds.), *The social amplification of risk* (pp. 326–352). Cambridge: Cambridge University Press.

Freudenburg, W. (1988). Perceived risk, real risk: social science and the art of probabilistic risk assessment. *Science, 242*, 44–49.

Freudenburg, W. (1993). Risk and recreancy weber, the division of labor, and the rationality of risk perceptions. *Social Forces, 71*(4), 909–932.

Freudenburg, W. (2003). Institutional failure and the organizational amplification of risks: The need for a closer look. In N. Pidgeon, R. E. Kasperson, & P. Slovic (Eds.), *The social amplification of risk* (pp. 102–120). Cambridge: Cambridge University Press.

Gadgil, M. (2014). Knowledge as power. *The Hindu*. 13 January.

Gaikwad, R. (2014). Under nuclear plant's shadow, Moticher's women refuse to vote. *The Hindu*. 28 April.

Gupta, C., & Sharma, M. (2008). *Contested coastlines: Fisherfolk, nations, and borders in South Asia*. New Delhi: Routledge.

Healy, S. (2004). A 'post foundational' interpretation of risk: Risk as 'performance'. *Journal of Risk Research, 7*(3), 277–296.

Jasanoff, S. (1994). *Learning from disaster: Risk management after Bhopal*. Philadelphia: University of Pennsylvania Press.
Jayaraman, N. (2013). The coast is not clear. *Hindustan Times*. 10 March.
Jebaraj, P. (2012). Environment reports have 'cooked data' says tribunal. *The Hindu*. 13 February.
Kasperson, J. X., Kasperson, R. E., Pidgeon, N. F., et al. (2010). The social amplification of risk-assessing 15 years of research and theory. In P. Slovic (Ed.), *The feeling of risk: New perspectives on risk perception* (pp. 317–344). London: Routledge.
Klinke, A., & Renn, O. (2014). Expertise and experience: A deliberative system of a functional division of labor for post-normal risk governance. *Innovation: The European Journal of Social Science Research, 27*(4), 442–465.
Law, J. (1994). *Organizing modernity*. Oxford: Blackwell.
Lockie, S., & Measham, T. (2012). Social perspectives on risk and uncertainty: Reconciling the spectacular and the mundane. In S. Lockie & T. Measham (Eds.), *Risk and social theory in environmental management* (pp. 1–14). Collingwood: CSIRO Publishing.
Muralidharan, S. (2004). Bhopal: Continuing crisis institutional. *Economic and Political Weekly, 39*(49), 5196–5198.
Namey, E., Guest, G., Thairu, L., & Johnson, L. (2008). Data reduction techniques for large qualitative data sets. In G. Guest & K. M. McQeen (Eds.), *Handbook for team-based qualitative research* (pp. 137–162). Plymouth: AltaMira Press.
NDTV. (2011a). DMK joins Jayalalithaa against Kudankulam nuclear plant. *NDTV*. 18 October.
NDTV. (2011b). Students of Tata Institute of Social Sciences protest against Jaitapur N-plant *NDTV*. 11 May.
Perrow, C. (1991). A society of organizations. *Theory and Society, 20*(6), 725–762.
Reiss, Jr A. J. (1992) The institutionalization of risk. In: L. Clarke & J.F. Short (Eds.), (pp. 299–308). Boulder: Westview Press Oxford.
Renn, O. (2006). *Risk Governance: Towards an integrative approach*. Geneva: International Risk Governance Council.
Renn, O. (2014). Towards a socio-ecological foundation for environmental risk research. In: S. Lockie, D.A. Sonnenfeld & D. Fisher (Eds.), *Routledge international handbook of social and environmental change* (pp. 207–220). Abingdon, Oxon, New York: Routledge.
Renn, O. (2015). Stakeholder and public involvement in risk Governance. *International Journal of Disaster Risk Science, 6*(1), 8–20.
Renn, O., & Klinke, A. (2012). Complexity uncertainty and ambiguity in inclusive risk governance. In S. Lockie & T. Measham (Eds.), *Risk and social theory in environmental management* (pp. 59–76). Collingwood: CSIRO Publishing.
Roy, R. (2010). Haripur nuclear power plant becomes bone of contention. *Business Standard*. 10 February.
Saldana, J. (2009). *The coding manual for qualitative researchers*. London: Sage Publications.
Short, J. F. (1984). The social fabric at risk: Toward the social transformation of risk analysis. *American Sociological Review, 49*(6), 711–725.
Short, J. F., & Clarke, L. (1992). *Organizations, uncertainty and risk*. Boulder: Westview Press, Oxford.
Slovic, P. (1987). Perception of risk. *Science, 236*(4799), 280.
Slovic, P., Fischhoff, B., & Lichtenstein, S. (1979). Rating the risk. *Environment, 21*(3), 14–29.
Sovacool, B. K., & Valentine, S. V. (2010). The socio-political economy of nuclear energy in China and India. *Energy, 35*(9), 3803–3813.
Spradley, J. P. (1979). *The ethnographic interview*. Fort Worth, TX: Harcourt Brace Jovanovich.
Srikant, P. (2009). *Koodankulam anti-nuclear movement: A struggle for alternative development?* Working paper no. 232. The Institute for Social and Economic Change, Bangalore.
Srikant, P. (2010). *Reconceptualising rights—A study of micro social movements in India* (Unpublished doctoral thesis). University of Mysore, Bangalore.
Times of India. (2013). Parties oppose Haripur N-plant. *Times of India*. 21 October.
Vaughan, D. (1996). *The challenger launch decision: Risky technology, culture, and deviance at NASA*. Chicago, London: University of Chicago Press.

Vaughan, D. (1999). The dark side of organizations: Mistake, misconduct, and disaster. *Annual Review of Sociology, 25*(1), 271–305.
Wong, C. M. L. (2015a). The mutable nature of risk and acceptability: A hybrid risk Governance framework. *Risk Analysis, 35*(11), 1969–1982.
Wong, C. M. L. (2015b). Organisational risk perception and transformations in India's nuclear establishment. *Journal of Risk Research, 18*(8), 1012–1029.
Wynne, B. (1992). Misunderstood misunderstanding: Social identities and public uptake of science. *Public Understanding of Science, 1,* 281–304.
Wynne, B. (1996). May the sheep safely graze? A reflexive view of the expert-ay knowledge divide. In S. Lash, B. Szerszynski, & B. Wynne (Eds.), *Risk, environment and modernity: Towards a new ecology* (pp. 44–83). London: Sage Publications Ltd.

Author Biography

Dr. Catherine Mei Ling Wong is a sociologist at the University of Luxembourg. Her work focuses on risk governance in the energy sector, climate policy and global cities. Dr. Wong also specialises in qualitative comparative research methods. Having done her first postdoc at The Cairns Institute, James Cook University on climate governance in Australia, China and the UK, Dr. Wong now works with Geographers at the University of Luxembourg on globalisation, risk and governance in global financial centres. Catherine has a Ph.D. in sociology from the Australian National University, and a Master in Global Studies from the University of Freiburg, Germany (Dr. Wong can be contacted at: catherine.wongml@gmail.com).

Chapter 7
Public Perceptions of Nuclear Energy in Asia After Fukushima Crisis

Kapil Patil

7.1 Introduction

The Fukushima nuclear crisis triggered by massive earthquake and tsunami that struck Japan in March 2011 significantly affected the public acceptance of nuclear energy around the world. The accident took place at a time when nuclear energy was witnessing highest levels of growth and favourable public opinion around the globe. By the end of 2010, the global nuclear energy generation reached a record peak of 2620 billion KWh with 441 operational nuclear power reactors in 30 countries and as many as 67 new reactors under construction in 13 countries (IAEA 2010). The nuclear meltdown at Fukushima Daiichi station, however, shook public confidence in safety of nuclear power plants and affected nuclear energy policies in many countries. The nuclear crisis intensified public protests and drove countries such as Germany, Italy and Switzerland to phase out nuclear power, while countries such as China, South Korea and India opted for comprehensive safety reviews of the operating nuclear power stations to allay public fears. In China, the State Council suspended approvals for the new reactors in order to revise safety standards in the existing fleet of reactors (Xinhua 2011). The safety concerns in the wake of Fukushima disaster had threatened to derail the ambitious plans for expanding the role of nuclear power in the Asian regional energy mix. However, the emerging trends in nuclear energy expansion in the continent indicate a continued recovery in public acceptance of nuclear energy as a clean energy source with Japan as the sole exception. Compared to Europe and North America, most new nuclear capacity addition continues steadily in the Asian region after the temporary drop caused due to Fukushima nuclear disaster. Currently, there are about 40 nuclear reactors under construction in various Asian countries and a plan for building at least 90 reactors in the next two decades (WNA 2016). Besides the

K. Patil (✉)
Indian Pugwash Society, New Delhi, India
e-mail: kapilpatil24@gmail.com

N. Janardhanan et al. (eds.), *Resurgence of Nuclear Power*,
https://doi.org/10.1007/978-981-10-5029-9_7

continuing expansion in existing nuclear energy producing countries, the region has also seen the emergence of several "new entrant" countries that are pursuing nuclear energy plans mostly unchanged after the *Fukushima* crisis. The exaggerated fears about safety of nuclear installations in the aftermath of Fukushima did not result in downsizing of plans for nuclear energy expansion in the Asian countries. The continuing interest in nuclear energy in most of the Asian countries derives primarily from the fact that nuclear power remains one of the preferred energy sources to address the dual challenge of energy security and climate change. Significant improvements made in nuclear safety since the Fukushima accident has enabled nuclear utilities to recover rapidly from the post-Fukushima gloom.

The events in Japan, however, has once again highlighted that nuclear accidents can significantly affect public perceptions of risk, which in turn can decrease political as well as popular support for new nuclear energy build-up. The negative perceptions of risks fuelled mostly by the exaggerated reporting of the events could adversely affect government policies, at least in the short term. To ensure steady expansion of nuclear energy infrastructure, it is essential that public confidence in nuclear power must be restored through appropriate risk communication measures and that people's perceptions of risk must be attenuated through long-term technical and policy institutional changes. In this backdrop, the paper seeks to understand the dynamics of public attitudes towards nuclear energy in the historical context and survey ongoing trends in nuclear energy development plans in the Asian region after the Fukushima nuclear accident. The paper is organised in the following sections. The first section seeks to summarise the elements of public attitudes towards nuclear energy as outlined in the extant literature on sociology of risks perceptions (Adam et al. 2000). The second section reviews public responses to Fukushima crisis while analysing the key drivers of nuclear energy growth in leading Asian countries. It will also discuss the shifts in public opinion vis-à-vis nuclear energy in the years following Fukushima accident. The final section of the paper will analyse the impact of Fukushima crisis on nuclear energy development policies in the leading nuclear energy producing countries in Asia. The concluding section will summarise main findings pertaining to public acceptance of nuclear power in the Asian region.

7.2 Public Perceptions and Nuclear Power in the Historical Context

Nuclear energy has traditionally remained a contested issue with respect to public opinion. Although policy-makers in many countries have considered nuclear power as an important alternative for reducing dependence on fossil fuels, the negative public attitudes have often impeded government policies for nuclear energy deployment (Glaser 2012). The public opposition to nuclear energy became a serious concern in the decades of the 1970s and 1980s which saw large public protests against the construction of new nuclear power plants in countries such USA and Germany. The organised public opposition was mainly driven by the association of nuclear power with fears of catastrophe and ecological concerns. This affected the large-scale expansion of nuclear power necessitated by the 1973 oil crisis that caused serious energy deficit and global economic slowdown (Joppke 1991). Opinion polls by Harris and Gallup in the mid-1970s showed a considerably low public support for construction of new nuclear plants in USA (Harris 1975; Gallup 1976). Subsequently, events such as the Three Mile Island accident (1979) in the USA and Chernobyl (1986) in Ukraine further aggravated the perceptions of risks vis-à-vis nuclear power, thus making new nuclear build-up politically unfeasible in several European countries. Especially, after the Chernobyl disaster, public opinion polls consistently recorded low popular acceptance of nuclear energy for several years (Eiser et al. 1989; Verplanken 1989). Although countries such as Britain and France pushed ahead with building new reactors after Chernobyl accident, opposition to nuclear power hardened in several European countries. In France, the accident at its Super Phenix fast breeder reactor caused considerable public uproar over nuclear safety, whereas in Sweden, concerns over safety and environmental impacts of nuclear accidents precipitated public opposition to nuclear energy (Bupp 1987).

Public attitudes, thus became central to implementing government policies for nuclear energy and managing adverse perceptions of risks among lay people, attracted immense policy and academic attention. The public perceptions of risks are known to be highly dynamic and shifting over time due to changes in host of socio-economic, political and cultural conditions (Irwin et al. 2000). Also, the absence of any "single" public opinion on nuclear energy, as Pidgeon (2013) has noted through extensive survey of British population, poses a unique problem for presenting uniform and coherent understanding on public perceptions of risks towards nuclear energy (Pidgeon et al. 2008). The studies and surveys on public perceptions of risks have, nevertheless, outlined a number of variables that help explain the general societal perceptions towards nuclear energy. These include factors such as knowledge about nuclear technologies, trust in nuclear authorities, ideological predispositions, political and cultural beliefs that contribute to forming societal perceptions towards nuclear energy (Slovic et al. 1991, 2000; Rosa et al. 2000). The studies also point out some of the key attitudinal differences towards nuclear energy among lay people. For instance, the perceptions of risks among lay people are known to be significantly severe than the risks perceived by the technical experts. Due to inherently complex nature of the nuclear reactor technology, the lay people are often unable to understand nuclear energy risks in proportion to other technological hazards faced by them in a day-to-day life. Slovic (1987) described the systematic differences in risk perception and termed nuclear power as a dreadful and unknown hazard in public mind relative to other types of hazards. Owing to such adverse risk perceptions, local communities are often found opposed to siting of nuclear installations in their immediate neighbourhood (Ramana 2011; Rosa 2005). On the contrary, at the existing nuclear power installations; population groups exhibit a more favourable attitude, thus indicating a marked difference in perceptions of economic benefits, health and environmental risks (Van Der Pligt et al. 1986).

Freudenburg and Davidson (1996) point out a stark gender division in perceptions of risks in which "women tend to be significantly more concerned about nuclear energy risks than men". The political affiliation of groups in a given population is also highlighted as an important element shaping people's attitude. An important survey by Costa-Font et al. (2008) found that risk perceptions of nuclear power is often "aligned", or "anchored", with political affiliation or political ideology irrespective of the knowledge of nuclear power among the social groups. Using survey data from a 2005 Eurobarometer of UK citizens, Costa-Font et al. (2008) claimed that attitudes towards nuclear energy are driven much more by an instinctive political decision rather than digesting scientific information and that individuals with centre or right political leanings appear less likely to be fairly opposed or totally opposed to nuclear energy generation than those with left-leaning political stances. Trust in the nuclear governance institutions is found also to be a critical factor for reducing perceived risk of nuclear power and together higher trust and lower risk perceptions can have overall positive attitudes towards nuclear power (Whitfield et al. 2009). Contrastingly, factors such as secrecy or lack of transparency in regulatory institutions often erode public trust and therefore become predictors of public acceptance of risk (Sjoberg 2001). Finally, knowledge about nuclear energy risks and benefits is a crucial determinant of public support. Important surveys have noted that people who feel informed about nuclear safety tend to perceive fewer risks than those who feel uninformed (OECD 2010). A similar information gap exists vis-a-vis climate change among large parts of the public who are found to be mostly unaware of (or choose not to believe) the potential benefit of nuclear energy to reduce the emissions of climate change-related carbon dioxide (OECD 2010). Enhancing public knowledge about safety related risks and benefits for minimising climate related hazards in a balanced manner can lead to increased levels of public support. A complex interplay of psychological, social, cultural, historical and political thus condition public attitudes and perceptions of risks towards nuclear power which cannot be easily modified by sporadic public information or educational campaigns (Slovic 2000).

Towards the beginig of 21st century, nuclear energy began witnessing a shift in lay risk perceptions. A number of public survey polls since 2000 have shown a more supportive public opinion towards nuclear energy in the overall energy mix (OECD 2010). A survey by Knight (2005, 2009) in UK noted positive attitude towards nuclear energy both among the general public and members of parliament (MPs). Grove-White et al. (2006) too noted the limited but potential for increased public support based on climate change concerns. The relatively safe and accident-free performance of nuclear industry for close to two decades since Chernobyl disaster has led to more favourable public attitudes towards nuclear energy in most OECD countries (Eurobarometer 2007). Also, with climate change emerging as a serious concern for most industrialised countries, nuclear energy emerged as an important option for climate-friendly growth (Clery 2005). Pidgeon et al. (2008), however, noted that proportion of the British public accepts nuclear power as a means of addressing the greater threat of climate change only "reluctantly". The Eurobarometer polls (2007, 2010) illustrated that support for energy production by nuclear power stations has grown significantly in the European

Union since winter 2005, and Europeans on average have rather "moderate" opinions about nuclear energy. However, a survey across the 18 countries by IAEA showed that majorities of citizens while support the continued use of existing nuclear reactors are not favourable to building new ones with regional exceptions such as USA, South Korea, Britain, India (IAEA-Globescan 2005). The Fukushima crisis in Japan yet again shifted global public opinion causing immediate and widespread public opposition to nuclear power and policy shifts in some countries. The following sections attempt to analyse the public perception challenges after Fukushima crisis in the Asian region, where large nuclear energy growth is expected to take place in the coming decades.

7.3 Public Reaction to Fukushima Crisis: Drivers and Trends

Unlike Chernobyl disaster that caused serious trans-boundary radioactive contamination and fuelled negative opinion in many countries, the Fukushima nuclear crisis had more varied impacts on public opinion. The radioactive contamination caused by the Fukushima was only about 10% of what caused by the Chernobyl accident (von Hippel 2011). However, in countries such as Germany, Italy and Switzerland, the crisis significantly affected public opinion. In Italy, the public referendum on nuclear energy registered widespread negative vote, while in Germany, the government decided to phase out operating nuclear reactors by 2022. In Belgium too, the new government vowed to phase out its nuclear plants by 2015 (Digges 2011). The polls conducted in the immediate aftermath of the crisis showed little support across the world for building new nuclear reactors. One of the first major surveys commissioned by the BBC to Globe Scan polled 23,231 people in 23 countries during July–September 2011 showed that in countries with nuclear programmes, people are significantly more opposed to building more reactors than they were in 2005 (BBC 2011). The BBC poll, however, noted fairly stable public opinion in countries such as UK and USA in terms of support to nuclear energy. Public opinion polls after the Fukushima accident suggest widespread scepticism among lay people about nuclear power (BBC 2011). In Asian region, the public reaction to Fukushima nuclear crisis was rather muted with the exception of Japan, where a strong public sentiment exists against nuclear weapons after the atomic bombing of Hiroshima and Nagasaki. In countries such as China, Taiwan, South Korea and India, the dip in public support to nuclear energy proved short-lived and lasted for not more than few months. The heightened concerns after the Fukushima crisis, nevertheless, caused a transitory slowdown in nuclear energy growth and caused uncertainty about long-term expansion of nuclear energy in the Asian countries.

7.3.1 Drivers of Nuclear Energy Development in Asia

The Asian region has witnessed significant growth in nuclear power capacity addition in recent decades and is considered to be important for the future of nuclear energy. By 2008, the region became the centre stage for nuclear energy growth, with greatest growth in nuclear generation expected to occur in countries such as China, Japan, South Korea and India. The growing interest in nuclear energy development was driven by mainly three important imperatives. Firstly, the region came to be accounted for world's largest energy demand growth which was projected to almost double by 2030 (ADB 2013). The higher economic and population pressures underpinned by the growing industrialisation and urbanisation significantly spiked the energy demand in emerging Asian economies. Providing stable and reliable supply of electricity to large populations thus has been one of the important drivers of nuclear energy development in Asia. The energy demand in People's Republic of China (PRC) is projected to increase from 2417.1 Mtoe in 2010 to 3098.0 Mtoe in 2020 and 4218.1 Mtoe in 2035, growing at an annual rate of 2.3% (ADB 2013). Similarly, in India the primary energy demand is projected to increase from 692.7 Mtoe in 2010 to 1441.6 Mtoe in 2035, growing at an annual rate of 3.0% (ADB 2013). To meet the current energy demand though fossil fuels alone has proven both inadequate and unsustainable in the backdrop of adverse climate change threats. Countries such as Japan, South Korea and Taiwan import over 90% of petroleum, natural gas and coal from overseas, whereas, among the emerging Asian economies such India and China, the governments have been wary of depleting coal reserves. In China, the coal reserves account for around 14% of global reserves, but its share of global coal consumption has increased at an alarming 47%, towards the end of 2010 (Winning 2010). In India too, the high-quality coal for thermal reactors has always been in short supply, and dependence on fossil fuels is expected to increase enormously in the coming decades (IEA 2015). To minimise high dependence coupled with geopolitical uncertainties' surrounding fossil fuel supplies has been one of the key drivers of nuclear energy programmes in both developed and developing economies in Asia (Hu and Woite 1993). As a result, nuclear energy programmes in Japan, South Korea and Taiwan have grown enormously contributing significantly to their overall energy generation, whereas countries such as China and India had set ambitious targets for increasing share on nuclear energy on the overall energy mix. Thirdly, the imperative of carbon reduction goals has further prodded countries in the region to diversify the energy mix with increasing emphasis on renewable and (e.g. wind, solar and tidal energy) and nuclear energy. Nuclear energy, being a clean and load-intensive energy source, has thus become an attractive option for countries in the region, and a large number of new countries such as Thailand, Vietnam, Indonesia, the Philippines, Singapore and Myanmar have expressed interest in nuclear energy as part of their energy diversification strategy.

7.3.2 Public Acceptance After Fukushima

The immediate public reaction to Fukushima in many Asian countries, however, called for an end to new nuclear build-up. The small but extremely vocal communities of NGOs and civil society groups spearheaded opposition to nuclear energy in many of the Asian countries. In Japan, particularly, the public reaction to Fukushima crisis was most severe leading the government to review country's nuclear energy policies. The government of Prime Minister Naoto Kan in May 2011 announced scrapping of previous plans for building 14 new reactors by 2030 (Fackler 2011). Owing to strong public opposition, the government of Kan in July 2011 declared that Japan should "reduce and eventually eliminate dependence on nuclear energy" as Fukushima had demonstrated dangers of nuclear technology. By 5 May 2012, Japan's all operating power reactors went offline for extensive safety reviews and checks and so far only five reactors began operation under new safety standards introduced following the March 2011 accident at the Fukushima Daiichi plant (WNN 2016a, b). Elsewhere in Asia, countries generating nuclear energy such as China, South Korea, Taiwan and India too witnessed public concerns causing temporary slowdown in their nuclear capacity addition. In China, the concerns over nuclear safety led to calls for stringent safety reviews of the operating reactor as well as those under construction. More importantly, the Chinese State Council suspended approvals for new nuclear plant construction in the immediate aftermath of Fukushima. In South Korea and Taiwan, the anti-nuclear sentiment grew significantly after with protest demonstrations held in Seoul and Taipei. In RoK, the popular support to country's nuclear sector suffered a setback with the Fukushima accident. The local communities at newly designated plant sites staged protests to oppose construction of nuclear reactors. The Kori Nuclear Plant, in particular, became a subject of nationwide debate due to geological similarities with Fukushima Daiichi plant as well as the vulnerability of densely populated regions which fall within a 30-kilometre radius of the plant. The protests in Yeongdeok and Samcheok also exerted pressures on the Korean government, caught in a fix to continue with nuclear as a major source of electricity amidst concerns over imported fossil fuel supplies (Bird 2012). In Taiwan, public perceptions of nuclear power changed dramatically following the Fukushima accident. Major protests in the spring of 2014 over Lungmen project led the Taiwanese government to mothball the plant and to hold public referendum to decide the fate of the project. With new government of Democratic Progressive Party in Taiwan vowing to phase out country's nuclear energy by the year 2025, the future of nuclear energy in Taiwan remains uncertain (Jenning 2016). In India, too, the Fukushima accident resulted in surge of anti-nuclear demonstrations, particularly, at new nuclear plant sites such as Jaitapur, where 6 European pressurised reactors (EPR) are being planned in collaboration with France, and Kudankulam, where the opposition from local communities delayed the commissioning of two 1000-MWe VVER reactors built with the Russian assistance. The opposition from local communities at new nuclear plant sites in India has been mostly driven by perceived loss of benefits to local people due to land acquisition and loss of livelihood, etc. (Sethi 2012).

7.3.3 Recovery in Public Acceptance

Notwithstanding the public protests and demonstrations led by civil society groups and NGOs, the perceptions of lay people about nuclear energy in the Asian region vary from country to country. Although there are few systematic country studies on public perceptions in the context of Asian countries, the public opinion polls and surveys in recent years indicate a significant recovery in public acceptance in a number of countries. A survey by Ipsos Mori, in 2012, indicated that overall support for nuclear energy was strikingly evident in developing nations with strong nuclear energy programmes, including India at 75% and China at 59% (NEI 2012). In Japan and Taiwan, however, regular polls since the disaster show that the majority of citizens oppose the use of nuclear energy. The country-specific divergences in public acceptance of nuclear power in the Asia, however, appear unlikely to impinge on the overall growth trajectory of nuclear power in the region. To this effect, the role played by strong centralised governments in several countries of Asia remains crucial. The governmental emphasis on nuclear power as a solution to reduce dependency on fossil fuels and to pursue climate-friendly energy alternatives has ensured recovery in public trust. Besides, the majority population in Asia's new nuclear entrants too are largely favourable to nuclear energy, at least in the near future. Secondly, the relatively safe operation of nuclear power plants has made it possible for nuclear utilities in countries such as China and India to recover from the post-Fukushima slowdown and ensure higher safety benchmarks in nuclear power generation. Third and most importantly, the favourable public opinion is also underpinned by the societal belief that benefits of nuclear power considerably outweigh the risks and has bolstered public support for sustained expansion of nuclear energy in the region. As a result, the opposition in the wake of Fukushima crisis has not significantly altered government policies for nuclear energy development. The next section thus discusses the impact of Fukushima crisis and public attitudes on nuclear energy policies in some of the leading Asian countries.

7.3.4 Impact of Fukushima and Nuclear Energy Expansion in Asia

As discussed earlier, the Fukushima accident in Japan was a reminder of previous nuclear disasters such as Three Miles Island and Chernobyl, prompting a debate over the safety of nuclear power generation around the world. Adding to the public anxiety was the exaggerated media coverage which compelled many countries to review their nuclear energy policies. Yet, the public debate over the wisdom of nuclear energy programme has done little to stall the expansion of nuclear energy projects in many Asian countries. Among the countries in Asia, India, China and South Korea account for the largest number of nuclear energy projects to meet its rapidly growing energy requirements. Although Japan's own nuclear energy programme witnessed a major setback with Fukushima, the ruling Liberal Democratic Party (LDP) in Japan remains strongly committed in its pursuit for nuclear power as a major source of generating electricity in the country (Smith 2012). Amidst strong anti-nuclear protests by a large section of the Japanese population, Japan's official position of restarting some of the operable reactors stands in sharp contrast to European countries such as Germany and Switzerland which decided to phase out nuclear plants in a time-bound manner. Thus, the factors demanding the shutdown of nuclear energy projects in the post-Fukushima years have hardly succeeded to downsize the momentum of atomic energy projects in Asia.

Having the largest number of nuclear reactors in Asia and several under construction, Beijing's initial reaction immediately after the Fukushima crisis was a cautionary one, with government tightening the nuclear safety rules and adjustment carried out in "National Nuclear Long-and-Medium Term Development Planning" adopted in the year 2007 (Wang and Chen 2012). A small section of the Chinese population expressed concerns about radioactive contamination, in case of accidents, and became sceptical about the country's nuclear projects. As a response to acute safety concerns, the State Council called for a comprehensive inspection of the safety of China's nuclear facilities, of the both operational as well as the under-construction ones (Ming et al. 2016). The central nuclear enterprises such as the China National Nuclear Corp (CNNC), China Guangdong Nuclear Power Group (CGN) and China Nuclear Engineering Group Corporation (NEC) conducted a quality and safety inspection of nuclear facilities and nuclear projects under construction. However, the Chinese initial reaction was soon replaced by a policy affirmation to carry out the expansion of nuclear energy. The Ministry of Environmental Protection declared that there will be no changes in China's determination to develop nuclear power programme (Fu 2011). China regards nuclear energy as the most promising form of energy to help alleviate the energy crisis, thus controlling environmental pollution and in mitigating climate change threats. Followed by an extended reflection phase, the Chinese Premier Wen Jiabao, in October 2012, announced that China would "steadily return to normal construction" of new nuclear power plants, based on a "steady advance in an orderly manner" (WNN 2016a, b). Under the 13th Five-Year Plan, China is expected to have some 58 GWe of nuclear generating capacity in operation by 2020, up from the current capacity of almost 27 GWe. In addition, a further 30 GWe of nuclear capacity will be under construction by 2020 (WNN 2016a, b). The figures, thus, clearly speak for themselves. It reasserts that nuclear energy remains a crucial component of China's sustainable development goal while addressing environmental concerns.

As a country deficit of energy resources, South Korea has a highly developed nuclear sector, contributing up to 26% of its electricity, second only to coal generation (Dalton and Cha 2016). To restore public confidence on nuclear programme after the Fukushima crisis, Seoul has opened the world's first International Nuclear Safety School to train safety experts from other countries (O'Donell 2013). Notwithstanding the anti-nuclear sentiment, the Korean government adheres to its original plan to continuously rely on nuclear energy and continuing with nuclear power plant projects mostly unchanged. Given the country's heavy dependence on energy imports, any plan to reverse the nuclear energy projects would cost the Korean economy heavily. Low-cost nuclear electricity has been a boon for industry and Korea's critical manufacturing sector. In addition to its domestic benefits, South Korea has major commercial interests in promoting nuclear power. Its first attempt to make a foray in the nuclear technology export market was the contract signed between a South Korean consortium led by KEPCO and the United Arab Emirates in December 2009 for four APR-1400 reactors, worth $20 billion (York 2013). It also hopes to win a share of the lucrative market for the maintenance and repair of reactors worldwide.

In India's case, the Fukushima crisis fuelled public opposition from local communities at various newly proposed nuclear power plant sites in the country. To assuage safety-related fears and concerns, the Nuclear Power Corporation Ltd (NPCIL) and the Atomic Energy Regulatory Board (AERB) have carried out several safety reviews and modifications to the existing fleet of nuclear reactors, and the country remains committed to secure its energy mix through ensuring sustained nuclear energy growth. The Fukushima crisis has in no significant way derailed the country's growing nuclear plans. In the wider Asia-Pacific region, a large number of countries expressed keen interest in developing nuclear energy as a low-carbon alternative source to their energy mix. Plans were formulated along with proposals set up for nuclear power reactors in countries such as Thailand, Malaysia, Indonesia, Philippines and Vietnam. The Fukushima accident, however, has merely delayed the plans to develop nuclear energy programme in these countries, and after an initial wait-and-see period followed by a cautionary halt, these countries have come to realise the importance of nuclear power in their energy strategies.

Amid fossil fuel shortages and substantial rises in energy demand, the nuclear energy has become more of a means than a choice for the countries of this region. Thailand currently envisages having a 2000 MW of nuclear capacity in line by 2026 (Lee et al. 2014). Despite challenges related to public acceptance of nuclear energy in Malaysia, the Malaysian Nuclear Power Corporation is all set to commission its first nuclear plant by 2023 (Khattak et al. 2016). Although technology selection and safety precautions delayed plans for Vietnam, it has signed nuclear cooperation agreements with the Russian Federation and Japan and plans to develop up to 10 nuclear plants by 2030 (Vivoda 2013). Thus, not many changes in government policies have happened with regard to nuclear energy development in this region. Notwithstanding the dip in public support to nuclear power in most Asian countries in the post-Fukushima period, the significance of nuclear power as an alternative to fossil fuels has only grown in the face of rising energy demands globally. The rational consideration of energy alternatives, after the post-Fukushima cautionary phase, has thus restored region's nuclear energy development to steady growth.

7.4 Conclusion

In the aftermath of Fukushima nuclear disaster, the public acceptance became a crucial impediment for the future growth of nuclear energy. Although the public concerns over nuclear safety in the wake of Fukushima did not result in reversal of government policies for nuclear energy development in many countries, the crisis has once again crisis called for a renewed commitment to nuclear safety coupled with the enhanced risk communication measures to manage lay perceptions of risks. The slow but steady recovery in public acceptance vis-à-vis nuclear energy in many countries is reinforced due to greater benefits associated with the use of nuclear power for meeting the goals of sustainable socio-economic development. However, the localised public opposition at nuclear plant sites in many countries has strained the relationship between society as a whole and the institutional structures that promote nuclear power. Addressing such local opposition and retaining larger societal trust still remains a key challenge for nuclear energy in post-Fukushima scenario. The proactive and engaged efforts such as building constructive public relations and reassurance about the benefits of nuclear energy use will be a key to regaining public confidence in nuclear energy.

References

Adam, B., Beck, U., & van Loon, J. (2000). *The risk society and beyond: Critical issues for social theory* (Vol. 3, pp. 211–230). Sage Publications.

Asian Development Bank. (2013). Energy outlook for Asia and the Pacific. October 2013. http://adb.org/sites/default/files/pub/2013/energy-outlook.pdf.

BBC News. (2011). Nuclear power 'gets little public support worldwide' by Richard Black, November 25, 2011. http://www.bbc.com/news/science-environment-15864806.

Bird, W. (2012). Anti-nuclear movement growing in Asia. *Christian Science Monitor*, January 27, 2012. http://www.csmonitor.com/World/Asia-Pacific/2012/0127/Anti-nuclear-movement-growing-in-Asia.

Bupp, I. C. (1987). *Chernobyl's fallout: The future of nuclear power*. Cambridge, Mass: Cambridge Energy Research Associates.

Clery, D. (2005). Nuclear industry dares to dream of a New Dawn. *Science*, *309*(5738), 1172–1175. Retrieved from http://science.sciencemag.org/content/309/5738/1172.abstract.

Costa-Font, J., Rudisill, C., & Mossialos, E. (2008). Attitudes as an expression of knowledge and "political anchoring": The case of nuclear power in the United Kingdom. *Risk Analysis: An Official Publication of the Society for Risk Analysis, 28*(5), 1273–1288. doi:10.1111/j.1539-6924.2008.01094.x.

Dalton, T., & Cha, M. (2016). South Korea's nuclear energy future. *The Diplomat*, February 23, 2016. http://thediplomat.com/2016/02/south-koreas-nuclear-energy-future/.

Davidson, D. J., & Freudenburg, W. R. (1996). Gender and environmental risk concerns: A review and analysis of available research. *Environment and Behavior, 28*(3), 302–339. doi:10.1177/0013916596283003.

Digges, C. (2011). Belgium plans conditional nuclear phase out, joining other European nations that are deserting nuclear power. Bellona. November 7, 2011. http://bellona.org/news/nuclear-issues/nuclear-meltdown-in-japan/2011-11-belgium-plans-conditional-nuclear-phase-out-joining-other-european-nations-that-are-deserting-nuclear-power.

Eiser, J. R., Spears, R., & Webley, P. (1989). Nuclear attitudes before and after chernobyl: Change and judgment1. *Journal of Applied Social Psychology, 19*(8), 689–700. doi:10.1111/j.1559-1816.1989.tb00348.x.

Eurobarometer. (2010). Europeans and Nuclear Safety Report. *Survey Co-Ordinated By Directorate General Communication.* http://ec.europa.eu/commfrontoffice/publicopinion/archives/ebs/ebs_324_en.pdf.

European Commission. (2007). Special Eurobarometer. *Europeans and Nuclear Safety Report, l* 271/Wave 66.2–TNS Opinion & Social. Available at: http://ec.europa.eu/public_opinion/archives/ebs/ebs_271_en.pdf.

Fackler, M. (2011). Japan scraps plan to build new reactors; Prime minister shelves reactor plans and vows new focus on renewables. *International Herald Tribune.* May 11, 2011. http://www.nytimes.com/2011/05/11/world/asia/11japan.html?_r=0.

Fu, W. (2011). Nuclear fallout. *Global Times*, March 15, 2011. http://www.globaltimes.cn/content/633233.shtml. Accessed August 23, 2016.

Gallup, G. (1976). Most See Need for Nuclear Power, But Many Concerned About Safety. Chicago: Field Newspaper Syndicate. July 22, 1976.

Glaser, A. (2012). From Brokdorf to Fukushima: The long journey to nuclear phase-out. *Bulletin of the Atomic Scientists, 68*(6), 10–21. doi:10.1177/0096340212464357.

Global Public Opinion on Nuclear Issues and the IAEA, Final Report from 18 Countries Prepared for The International Atomic Energy Agency by GlobeScan Incorporated October 2005.

Grove-White, R., Kearnes, M., Macnaghten, P., & Wynne, B. (2006). Nuclear futures: Assessing public attitudes to new nuclear power. *The Political Quarterly, 77*(2), 238–246. doi:10.1111/j.1467-923X.2006.00766.x.

Harris, L., & Associates. (1975). *A Survey of public and leadership attitudes toward nuclear power development in the United States*. New York: Ebasco Services.

Hu, C., & Woite, G. (1993). Nuclear power development in Asia. *IAEA Bulletin*, 4/1993. https://www.iaea.org/sites/default/files/publications/magazines/bulletin/bull35-4/35404780207.pdf.

International Atomic Energy Agency. (2011). International Status and Prospects of Nuclear Power (2010 Edition). March 2011. https://www.iaea.org/sites/default/files/np10.pdf.

Irwin, A., Allan, S. & Welsh, I. (2000). Nuclear risks: Three problematics. In: B. Adam, U. Beck & J. van Loon (Eds.), *The risk society and beyond: Critical issues for social theory* (pp. 79–104). London: Sage Publications Ltd. doi:10.4135/9781446219539.n4.

Jenning, R. (2016). Taiwan nuclear power shutdown 'looks unsustainable'. Nikkaei Asian Review. July 30, 2016. http://asia.nikkei.com/Politics-Economy/Economy/Taiwan-nuclear-power-shutdown-looks-unsustainable, Accessed August 21, 2016.

Joppke, C. (1991). Social movements during cycles of issue attention: The decline of the anti-nuclear energy movements in West Germany and the USA. *The British Journal of Sociology, 42*(1), 43–60. doi:10.2307/590834.

Khattak, M. A., Arifb, A. A., Hannanc, A., Syukrid, F., & Hamid, H. (2016). Design and planning of a nuclear power plant in Malaysia: A feasibility report. *Journal of Advanced Research in Applied Sciences and Engineering Technology, 3*(1), 67–76.

Knight, R. (2005). What the polls tell us. Nuclear Engineering International, April 1: 24–5.

Knight, R. (2009). *Public attitudes to the nuclear industry*. London: Ipsos MORI.

Lee, M., Park, D., & Saunders, H. (2014). Asia's Energy Adequacy, Environmental Sustainability, and Affordability: An Overview, Asian Development Bank, No. 398, June 2014. http://www.adb.org/sites/default/files/publication/42605/ewp-398.pdf.

Ming, Z., Yingxin, L., Shaojie, O., Hui, S., & Chunxue, L. (2016). Nuclear energy in the Post-Fukushima Era: Research on the developments of the Chinese and worldwide nuclear power industries. *Renewable and Sustainable Energy Reviews, 58,* 147–156. doi:10.1016/j.rser.2015.12.165.

Nuclear Energy Institute. (2012). Global Poll Shows Nuclear Support Increasing After Fukushima. October 17, 2012. http://safetyfirst.nei.org/news/global-poll-shows-nuclear-support-increasing-after-fukushima/.

O'Donnell, J. K. (2013), Nuclear Power in South Korea's Green Growth Strategy: Green Growth Quarterly Update III-2013, Council for Foreign Relations. June 2013. http://www.cfr.org/south-korea/nuclear-power-south-koreas-green-growth-strategy/p31030.

OECD Nuclear Energy Agency. (2010). Public attitudes to nuclear power. NEA No. 6859. Paris: NEA.

OECD/IEA. (2015). India Outlook Energy–World Energy Outlook Special Report. http://www.worldenergyoutlook.org/media/weowebsite/2015/IndiaEnergyOutlook_WEO2015.pdf.

Pidgeon, N. F., Lorenzoni, I., & Poortinga, W. (2008). Climate change or nuclear power—No thanks! A quantitative study of public perceptions and risk framing in Britain. *Global Environmental Change, 18*(1), 69–85. doi:10.1016/j.gloenvcha.2007.09.005.

Poortinga, W., Aoyagi, M., & Pidgeon, N. F. (2013). Public perceptions of climate change and energy futures before and after the Fukushima accident: A comparison between Britain and Japan. *Energy Policy, 62,* 1204–1211. doi:10.1016/j.enpol.2013.08.015.

Ramana, M. V. (2011). Nuclear power and the public. *Bulletin of the Atomic Scientists, 67*(4), 43–51. doi:10.1177/0096340211413358.

Rosa, E. A. (2005). Déjà Vu all over again for nuclear power? *Science, 310*(5748), 619. Retrieved from http://science.sciencemag.org/content/310/5748/619.4.abstract.

Rosa, E. A., Matsuda, N., & Kleinhesselink, R. R. (2000). The cognitive architecture of risk: Pan-cultural unity or cultural shaping? In O. Renn & B. Rohrmann (Eds.), *Cross-cultural risk perception* (pp. 187–207). Dordrecht: Kluwer.

Sethi, M. (2012). Future of nuclear power in India. *Claws Journal.* Summer 2012. http://www.claws.in/images/journals_doc/1394791368Manpreet%20Sethi%20%20CJ%20Sumer%202012.pdf.

Sjoberg, L. (2001). Limits of knowledge and the limited importance of trust. *Risk Analysis, 21*(1), 189–198. doi:10.1111/0272-4332.211101.

Slovic, P. (1987). Perception of risk. *Science 236*(17 April), 280–285.

Slovic, P., Flynn, J. H., & Layman, M. (1991). Perceived risk, trust, and the politics of nuclear waste. *Science, 254*(5038), 1603–1607. Retrieved from http://science.sciencemag.org/content/254/5038/1603.abstract.

Slovic, P., Flynn, J., Mertz, C. K., Poumadère, M., & Mays, C. (2000). Nuclear power and the public: A comparative study of risk perception in France and the United States. In O. Renn & B. Rohrmann (Eds.), *Crosscultural risk perception* (pp. 55–102). Dordrecht: Kluwer.

Smith, S. A. (2012). Anti-nuclear sentiment and Japan's energy choices. *Council for Foreign Relations.* April 12, 2016. http://blogs.cfr.org/asia/2016/04/12/anti-nuclear-sentiment-and-japans-energy-choices/.

Tabuchi, H. (2011). Japan Premier Wants Shift Away From Nuclear Power. New York Times, July 13, 2011. http://www.nytimes.com/2011/07/14/world/asia/14japan.html.

Van Der Pligt, J., Eiser, J. R., & Spears, R. (1986). Attitudes toward nuclear energy: Familiarity and salience. *Environment and Behavior, 18*(1), 75–93. doi:10.1177/0013916586181004.

Verplanken, B. (1989). Beliefs, attitudes, and intentions toward nuclear energy before and after chernobyl in a longitudinal within-subjects design. *Environment and Behavior, 21*(4), 371–392. doi:10.1177/0013916589214001.

Vivoda, V. (2013). After Fukushima: The future of nuclear power in Asia. Internet: www.eastasiaforum.org/2013/03/16/after-fukushima-the-future-of-nuclearpower-in-asia.

Von Hippel, F. N. (2011). The radiological and psychological consequences of the Fukushima Daiichi accident. *Bulletin of the Atomic Scientists, 67*(5), 27–36.

Xinhua. (2011). China suspends approvals for new nuclear plants. March 16, 2011. http://news.xinhuanet.com/english2010/china/2011-03/16/c_13782113.htm.
Wang, Q., & Chen, X. (2012). Regulatory failures for nuclear safety—the bad example of Japan—implication for the rest of world. *Renewable and Sustainable Energy Reviews, 16*(5), 2610–2617. doi:10.1016/j.rser.2012.01.033.
Whitfield, S. C., Rosa, E. A., Dan, A., & Dietz, T. (2009). The future of nuclear power: Value orientations and risk perception. *Risk Analysis: An Official Publication of the Society for Risk Analysis, 29*(3), 425–437. doi:10.1111/j.1539-6924.2008.01155.x.
Winning, D. (2010). China's coal crisis. *Wall Street Journal*. November 16, 2010. http://www.wsj.com/articles/SB10001424052748704312504575617810380509880.
World Nuclear Association. (2016). Asia's Nuclear Energy Growth. Last Updated January 2016. http://www.world-nuclear.org/information-library/country-profiles/others/asias-nuclear-energy-growth.aspx.
World Nuclear News (2016a) Fifth Japanese power reactor restarted. August 12, 2016.http://www.world-nuclear-news.org/C-Fifth-Japanese-power-reactor-restarted-1208164.html.
World Nuclear News. (2016b). Nuclear growth revealed in China's new Five-Year Plan. March 23, 2016. http://www.world-nuclear-news.org/np-nuclear-plans-revealed-in-chinas-new-five-year-plan-2303166.html.
York, J. (2013). The Future of Nuclear Power in South Korea after Fukushima. E-International Relations. April 11, 2013. http://www.e-ir.info/2013/04/11/the-future-of-nuclear-power-in-south-korea-after-fukushima/. Accessed September 02, 2016.

Author Biography

Mr. Kapil Patil is a researcher at the Indian Pugwash Society, New Delhi. His research focuses on issues of nuclear energy, non-proliferation and disarmament. He is currently involved in a research study on India's nuclear industry. Previously, he worked on a Department of Atomic Energy Sponsored research project and is a co-author of the book titled, "India's Nuclear Energy Programme: Challenges, prospects, and public concerns" published in 2014. He is also affiliated to Nuclear Proliferation International History Project (NPIHP) and participated in the Nuclear History Boot Camp in May 2011. His research has featured in Global Energy Monitor, Nautilus, The Diplomat, CBW Magazine, etc. (Mr. Kapil can be contacted at: kapilpatil24@gmail.com).

Part III
Knowledge Management

Chapter 8
Nuclear Knowledge Management—Trends and Good Practices for Asia

Yanko Yanev and Tatiana Karseka

8.1 Background

The post-Chernobyl decline in the support for nuclear energy toward the end of the twentieth century had a major impact on the nuclear industry as a whole and, in particular, to its access to a highly qualified and experienced workforce. Then, the first decade of the twenty-first century saw a modest renaissance in the application of nuclear technology, driven by (i) the need for diversification of energy technologies to mitigate the effects of global climate change, (ii) the need arising from the rapid growth of new economies, and (iii) the increasing demand for raw materials and minerals. The potential benefits of nuclear applications including nuclear power were, once again, being brought to the attention of expert communities and public around the world—until March 11, 2011.

The colossal natural disaster that lashed the east coast of Japan destroyed not only the Fukushima reactors but also the recently restored public confidence in nuclear energy in many countries in Europe, Asia, and America. Although tens of thousands were killed by the earthquake and the tsunami, not one person died from an overdose of radiation. Yet, the strong reaction from the public all over the world, spurred by media misinformation and a lack of timely factual information, has unleashed a renewed anti-nuclear crusade. Old fears have been compounded by a

---

The authors have been working on the area of nuclear knowledge management for several years. The paper uses certain contents from the published materials of the international atomic energy agency (IAEA) with consent.

Y. Yanev (✉)
Nuclear Knowledge Management Institute, Vienna, Austria
e-mail: y.yanev@nuclear-km.org

T. Karseka
International Atomic Energy Agency (IAEA), Vienna, Austria
e-mail: t.karseka@iaea.org

N. Janardhanan et al. (eds.), *Resurgence of Nuclear Power*,
https://doi.org/10.1007/978-981-10-5029-9_8

new one, stemming from the fact that a serious nuclear accident has occurred in a technologically advanced country that has stringent nuclear safety requirements. If the Japanese nuclear experts could not prevent it, then, according to public perception, reactors worldwide are all vulnerable to accidents, waiting to happen.

Fukushima nuclear power accident has brutally confirmed the vitality of effective nuclear knowledge management and coordination through the lack of it. It was critically important to have such a set up that in emergency, information reaches the right people, in the right place and at the right time. However, not all needed knowledge was available on the spot. Neither was all critical knowledge transferred and protected during the generation change or staff rotations. This intensified doubts concerning credibility of nuclear industry in society at large (IAEA 2016a, b).

To understand the relevance of nuclear knowledge management not only for emergencies like Fukushima but for all nuclear activities, it should be viewed in historical perspective. The international atomic energy agency (IAEA) has been a repository for nuclear data from the time the organization was established in 1957, but the subject of nuclear knowledge management came to the forefront formally only in the twenty-first century. The key concept is based on the assumption that knowledge, competence, and experience are the nuclear industry's most valuable asset, without which safe operation might be compromised, causing eventual exit.

8.2 What Is Nuclear Knowledge Management?

In course of last decades knowledge management became an area, which has received special attention in the nuclear industry especially in nuclear operating organizations but as well in organizations engaged in regulatory, research, and development activities. International organizations such as IAEA, NEA, and WANO have provided guidance and support to countries and organizations to implement knowledge management programs in nuclear activities. To define the scope of knowledge management in nuclear field, the IAEA clearly defined the most crucial terms which are important to understand the scope of the trends and efforts (IAEA 2012):

Nuclear knowledge management is the integrated, systematic approach applied to all stages of the nuclear knowledge cycle, including its identification, sharing, protection, dissemination, preservation and transfer, relevant to achieving the objectives of the nuclear energy program.

Knowledge is the vital resource for human existence and development. It must be managed consciously. Within the context of nuclear technology, careful and systematic approach is even more important, as nearly every discipline of physical sciences, social sciences, engineering, and economics is involved.

Nuclear knowledge is unique in many ways: it is complex, requires significant financial commitment and government support and must be developed, shared and transferred over many generations. Furthermore, there are commercial and security sensitivities with some critical information because of dual use (civilian and military) applications. Finally, the combination of personal skills and experience needed to turn this information into useable knowledge is particularly vital but rare. Levels of nuclear knowledge can be classified in the nuclear industry as *know-what, know-how, and know-why*. The *know-what* level—implying sufficient understanding of the technology—is the minimum required for senior managers who make decisions and allocate resources. At the *know-how* level, experts have the technical capability to apply their knowledge to nuclear technology and systems. At the *know-why* level, experts have the fundamental understanding of basic phenomena required to create new nuclear knowledge. Each level presents its own series of challenges for stakeholders.

Nuclear knowledge stakeholders are "governments, regulators, operators, suppliers, consultants, R&D organizations, educational and training institutions, non-governmental organizations (NGOs), the general public, and environmental campaigners—each of them with different knowledge and information requirements" (IAEA 2012a, b).

Business functions include business strategy, safety management, human resource management (including competence development), information technology and communications, quality management, information and records management, risk management, and public relations. Each has different contributions to an organization's management of knowledge.

An effective nuclear knowledge management system is part of the overall management system and focuses on three primary elements: **people, processes, and tools**, operating within an **organizational culture** that recognizes the value and role of nuclear knowledge (IAEA 2006).

The creation of nuclear knowledge, its dissemination, sharing, transfer, and application can be achieved **only by people** who have appropriate competence, experience, and motivation. Working arrangements will often be controlled through the use of documented processes to ensure that activities are undertaken in an orderly and consistent way. This includes the control of knowledge—explicit which is documented and implicit which resides in the heads of employees.

The second element is the **processes**. In such a highly regulated environment as the nuclear industry, clear directives, governance, and oversight are essential to preserve safety, control risk and ensure efficiency. Through the adoption of agreed processes, the management of knowledge and quality of the work can be best controlled and monitored.

The final element is **tools**, i.e., **technology**. In the current information-loaded environment of the nuclear industry, computers and associated equipment will inevitably be used to help create and exploit knowledge by the true custodians of nuclear knowledge—people.

An organization needs to develop and implement a **strategy** and planning to provide a framework for applying knowledge management principles in the workplace. Clearly defined objectives, as well as quantitative and qualitative characteristics of both targeted outcomes and resources required will help establish the role of knowledge management in the nuclear organization.

8.3 Characteristics of Nuclear Knowledge

Approaches relevant to managing knowledge depend on the nature of knowledge. That is why it is important to understand the specifics of nuclear knowledge. In accordance to the IAEA, there are five characteristics to be considered when managing nuclear knowledge (IAEA 2012):

- **Complexity**—nuclear knowledge is highly complex on both the micro- and macro-scale. It is multidisciplinary and requires deep knowledge in areas, such as physical, chemical, radiological, and biological interactions of material. Besides that, because of its military origins, sociological, economic, political, and security aspects must all be considered as a whole.
- **Cost**—largely due to its complexity, the creation and maintaining of nuclear knowledge is very costly. Nuclear facility construction and operation are often large, complex engineering projects requiring sophisticated safety systems and well-trained staff. Research and fundamental education in nuclear science demands facilities which in terms of cash flow would never pay back resources investments in its construction.
- **Timescale**—the period between the creation of knowledge and its use can be very long. Time between a decision to construct an NPP and its commissioning can vary between 5 and 20 years; the time between radioactive waste being created and its disposal could be many decades. During this time, information and knowledge, and the ability to retrieve and interpret it must be retained.
- **Cooperation**—due to its complexity, many individuals, organizations, and even States usually work together and make use, as well as enrich nuclear knowledge available. The information and data used and the experience, skills, and insights applied must be carefully controlled and monitored to guarantee robustness. This requires existence of cooperation, process management, and quality assurance rules which will be followed by all parties involved.
- **Education**—as nuclear knowledge is complex and multidisciplinary, a solid knowledge base is required to be able to understand existing set up, and even more fundamental education is required to be able to innovate and apply nuclear science in a new practicable way.
- **Safety** is the first priority for operating in the nuclear industry. "Decisions affecting safety must be made using the best knowledge and information available. If they are not, the risk of causing harm or inconvenience increases. Good nuclear knowledge management is a critical enabler of safety, security, environmental and financial management" (IAEA 2012).

Nuclear knowledge management is at the heart of safety culture. Once in place, nuclear knowledge systems must be sustained and further strengthened. Complacency is the enemy of a robust knowledge system. It always has to be managed.

The **industrial infrastructure** required to create and maintain the full scope of nuclear knowledge can represent a significant economic and technical burden for many countries. This infrastructure can include laboratories to handle a wide variety of nuclear material, research and power reactors, hot cells, reprocessing plants, demonstration facilities, and disposal sites. These have to be supported by skilled operators, health physicists, regulatory and licensing bodies, and quality and financial controllers. In recent years, the cost of maintaining such an infrastructure has risen. Willingness to share nuclear knowledge means that more countries, particularly in developing regions, are likely to have access to life-changing technology without establishing the costly local infrastructure at the beginning and in the learning phase.

Effective management of nuclear facilities requires **suitably qualified personnel**. An important element of human resource management is the management of knowledge—the knowledge needed to perform assigned tasks competently and the additional knowledge acquired in carrying out those tasks. This knowledge will be needed by several generations of the workforce during the lifetime of a nuclear energy program. As the nuclear workforce ages and retires, the number of suitably qualified and experienced staff will decline and their knowledge will be lost. Action is being taken to address this with the development of **higher education programs** that focus on nuclear technology and its application. However, many of these programs are still in their infancy and rely on concerted government and academic support—and both bodies are under constant pressure to demonstrate that they make best use of public funding. It could take decades of support before the benefits are realized, and in the meantime, nuclear safety and security may be at risk and innovation compromised. **Specialist knowledge** is needed to apply nuclear technology in medicine, agriculture, industry, disease prevention, water management, electricity production, and mineral exploration. If the accumulated knowledge is lost, nuclear applications may stall and many generations could have a less secure and sustainable future.

The **need to access knowledge** created as a result of historical processes and activities will increase in the future to support continued safe operation of nuclear facilities and their eventual decommissioning and waste management. As new facilities are designed and constructed, it will be increasingly important to know and heed what has worked well in the past. This is particularly relevant to developing countries planning to exploit nuclear applications in any area.

8.4 Knowledge Management in Nuclear Organizations

Data, information, and knowledge critical to the operation of a nuclear facility are generated from the initial phases of research and development across the facility's lifecycle including its decommissioning. As an example, the fundamental engineering principles incorporated into the design of a NPP constitute essential knowledge for subsequent considerations in the licensing process as well as in considering later changes to the design basis to support facility modifications. Another obvious example is the accumulation of operating experience for use in developing 'lessons learned' and conducting training of new and incumbent personnel. Also, the importance of radiological exposure data in relationship to the optimal use of personnel in both routine operations and abnormal situations constitutes a critical body of knowledge.

It is important to note that an organization's focus and KM priorities may be quite different depending on where the organization is in this lifecycle. Some KM methods and techniques may stay essentially constant while others may vary considerably. As noted earlier, the people, processes, and technology involved will determine the KM needs and methods.

KM is the process through which organizations generate value from their intellectual and knowledge-based assets. As with all such concepts, the role of the leaders of an organization cannot be overstated. The tone and level of expectations set by the most senior manager of an organization will drive both the implementation and the results. Knowledge management is a vital component of change management. As KM initiatives are undertaken or enhanced, it is imperative that expectations and the reasoning behind those expectations are clearly communicated throughout the organization. And, as made reference to earlier in this document, a spirit of knowledge sharing must pervade the organization if the full potential of KM is to be realized. Sensitivity to the need for continual, consistent KM must become ingrained in the very fabric—the culture—of an organization if its benefits are to be achieved. Its practices must become a 'way of life'—not just a temporary, passing management fad. Knowledge management must be integrated into strategic planning; analysis and decision making; implementation of plans; and evaluation of results. This is why KM is vital to an integrated management system and is advocated by the IAEA for protecting people and the environment.

Many nuclear organizations have programmes in place to capture, store, and retrieve much, if not all, of the information/knowledge identified above. Therefore, implementing a KM programme is not about discarding existing practices; rather, applying a KM approach should improve the value of existing programmes to the organization through:

- Identifying **business, operational, and safety risks** due to knowledge gaps (such as knowledge loss risk assessment to identify where the organization is most at risk of loosing mission-critical knowledge);
- Increasing the **value of existing knowledge** (through mechanisms such as making it easier to retrieve knowledge/information when it is needed in the workplace and potential integration of data systems/banks);
- Converting **tacit knowledge to explicit knowledge** (where the value to the organization justifies the resources needed for this effort);
- Improving **operational and safety performance** through creating both new knowledge and better access to existing knowledge;
- **Continually learning** in a smart and lasting way from successes and failures at the individual, team, and organizational levels;

- **Effective transfer of knowledge** from an aging workforce to younger workers; and
- **Improved strategic planning and decision making** resulting from access to more and more reliable knowledge/information.

Managing knowledge effectively requires understanding of and attention to the concept of organizational knowledge rather than just the traditional notion of individual-centered knowledge. This shift can be addressed through the utilization of organizational core competencies that have proven themselves to be of value within many nuclear organizations.

The **management system** is recommended to promote and support nuclear knowledge management as a primary opportunity for achieving competitive advantage and maintaining a high level of safety. This approach ensures that organizations are able to demonstrate their long-term competitiveness and sustainability through actively managing their information and knowledge as a strategic resource that supports the establishment and maintenance of safe, high-level organizational performance.

The knowledge management process within organization is to be addressed and improved. The knowledge management process normally includes knowledge identification; knowledge acquisition and development; knowledge dissemination and use; and knowledge preservation.

Nuclear knowledge management
Effective nuclear knowledge management program should: • support the safe operation of all nuclear facilities • achieve gains in economic and operational performance • facilitate the inter-organizational and intergeneration • transfer of knowledge • encourage innovation and cooperation • improve security and responsible use of information.
In order to meet these objectives, nuclear knowledge management should be integrated with other business activities and operations, at all levels, as part of: • all large nuclear projects • management and governance of all organizations involved in research, development, and utilization of nuclear energy and radiation technologies • national (government) nuclear development plans and policies.
Basic ingredients of a nuclear knowledge management culture: • culture that promotes knowledge sharing • responsible, no-blame environment (reporting incidents/events as a learning opportunity and sharing lessons learned) • sharing rewarded environment • sharing knowledge methods and tools in place • leadership/commitment from the highest levels in the organization • clearly defined objective/problem to be tackled and the competencies needed.

8.5 Knowledge Management and Nuclear Safety Culture

Many countries have now recognized the importance of developing both business and safety culture where nuclear knowledge is managed as a resource and pre-requisite. This requires a holistic approach where stakeholders are empowered to make contributions to technical and contextual information and the tools and processes used to protect, share, and transfer the associated knowledge. In nuclear organizations, Safety Culture is usually defined as organization's values and behaviors—modeled by its leaders and internalized by its members—that serve to make nuclear safety the overriding priority (IAEA 1991).

Nuclear knowledge management is a key factor in ensuring the safe operation of existing nuclear facilities, and the effective application of current knowledge to the design and construction of new nuclear projects, and development of innovative approaches for the nuclear industry.

The knowledge in the organization should be managed as a resource (IAEA 2016a, b).

An effective knowledge management system which combines three primary elements (knowledge carriers): people and their expertise and experience; organizational processes and technology which involves substantial design-based knowledge, operating within an organizational culture that recognizes the value of nuclear knowledge, is paramount in the nuclear industry. **Decisions affecting safety must be made using the best knowledge and information available**. If they are not, the risk of causing harm or inconvenience increases. Good knowledge management is a critical enabler of safety, security, environmental, and financial management. Once in place, **knowledge management must be sustained and further strengthened**. It becomes a robust component of a working environment, and with appropriate support from the executive level, it turns to an organizational routine. In other words, safety and knowledge are interrelated and both are maintained with the means of culture and management practices, as shown in Fig. 8.1 (Karseka and Yanev 2013) (Fig. 8.2).

Nuclear safety is based on knowledge and experience. Managing organizational knowledge through establishing a skilled, competent environment for sharing knowledge and good practices is a key element in creating and maintaining safety culture. Recent experience of a number of nuclear organizations that was shared through the IAEA demonstrates that these challenges can be successfully addressed by a well-formulated knowledge management strategy and clear implementation plan which includes methods, tools, and techniques for assuring staffs' understanding and knowing of safety implication of every decision or action. As a result, a successfully implemented knowledge management strategy ensures comprehensive integration of critical safety-related and knowledge-based practices into organizational routines.

A special attention to knowledge management should be paid at nuclear regulatory authorities. Their fundamental objective is to ensure that nuclear utilities operate their plants at all times in an acceptably safe manner. On one hand, a

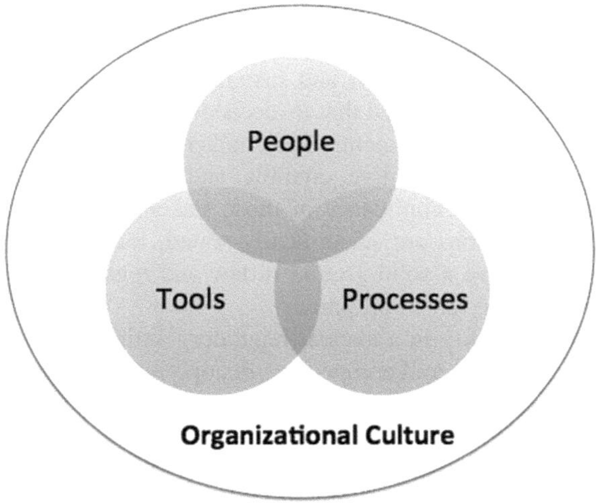

Fig. 8.1 Organizational structure

Fig. 8.2 Interfaces between safety culture and knowledge management

regulatory authority formulates requirements (prescriptive or performance based) and applies mechanisms to assure compliance (e.g., licensing process, inspections, or audits); on the other hand, it is not operating a power plant itself or conducting modifications. Maintaining competence and knowledge base of a regulatory authority is a critical but challenging task. In performing its primary objective, the regulatory authority should ensure that its regulatory decisions are technically sound, consistent from case to case, and sustainable in time. In addition, the regulator must be continuously alerted that its decisions in their context can affect how its stakeholders, such as government policy makers, the industry it regulates, and the public perceive it as an effective, credible, and independent regulator.

In order to maintain the confidence of those stakeholders, the regulator should make sure that its decisions are scientifically sound, technically correct and sustainable, transparent, have a solid ground in law and regulations, and are seen by impartial observers to be fair to all parties.

Knowledge sustainability in a nuclear regulatory authority is one of the critical factors that influence the overall competence in applying the regulatory process and forms the basis of a nuclear safety culture. Based on the accumulated experience with the current generation of nuclear power plants, it is considered that they will operate far beyond the originally design lifetime, involving change of generations, change of knowledge technology, and change of culture and attitude. In order to insure that relevant knowledge and skills will be available at any moment of the operational life of the nuclear installation, which can be more than 60 years, an integrated and systematic knowledge management program is needed in order to assure sustainability of the knowledge resource. Knowledge sustainability should be considered in overall organizational settings and include the following processes and activities:

- Knowledge management processes,
- Continuous learning,
- Managing the knowledge of the design rationale,
- Managing outsourced technical support services,
- Managing culture and attitude,
- Managing technical and regulatory changes,
- Managing design basis, and
- Knowledge reporting.

To ensure knowledge sustainability, it is recommended to establish the knowledge management process describing the necessary actions to fulfill the requirements of the knowledge management policy and of the international standards such as IAEA and ISO. The knowledge management process consists mainly of two main parts. The first part focuses on determining/identifying and making available the knowledge resources (human, structural, and relationship knowledge resources) required to operate the processes. After identifying the knowledge requirements, knowledge resources needed are to be acquired from within organization or from outside sources of competence. After the process implementation, it is necessary to review experience and learnings obtained from the implementation and performs capturing and transferring a new knowledge for the future use.

8.6 Relevant Challenges and Trends for Nuclear Knowledge Management in Asia

Knowledge management trends and challenges are different for different stages of a nuclear power program. Symbolically, we shall describe the specifics of the pre-construction stage of a nuclear power program, construction, operations, and the post-operational shut down or decommissioning stage. The pre-construction stage includes the pre-project activities including decision making, based on the IAEA guidance for a national infrastructure development in so-called new-comer countries (IAEA 2007). Asia is a region where all of the mentioned stages are present. This makes knowledge management a very multidimensional problem.

Since 2015, eight NPP constructions started worldwide, out of which six started in China. At the same time, eight other reactors in that country started operations (Schneider et al. 2016) in the same year. Under these circumstances of rapid development, knowledge management trends related to construction and operation are highly relevant for that country. India also has reactors in construction, operation, and permanent shut down. South Korea operates 25 reactors which provide about one-third of South Korea's electricity. Besides that, Korea is constructing NPP units in other countries, e.g., UAE. These three countries that use nuclear power technology have similar objectives in terms of nuclear knowledge management. The **management of risks**, associated with the operation of nuclear facilities and use of nuclear material, is at the forefront of counties priorities. Knowledge management has a vital role to play in risk management. To accurately assess risk, operators must have access to the right information at the right time and should have the deep knowledge and understanding of operational procedures and the competences and skills necessary to make **sound judgments and decisions based on credible information**. Furthermore, after-action reviews and 'lessons learned' exercises, which are important knowledge management processes, can be used to improve performance and efficiency and reduce future risks. Integrating knowledge management approaches in established quality management systems and, further on, in an integrated management system can result in greater control and improved performance of nuclear operations. For countries such as China, India, and Korea, this is one of the common trends and developments for every nuclear facility. At the national level, the operator, designer, regulator, and technical support organization will require the same knowledge, so the national authorities must play a role in facilitating inter-organizational knowledge transfer. Developing and managing this inter-organizational knowledge flow is a knowledge management trend and challenge which has to be addressed, especially now when the industry is on the brink of a new generation of nuclear workers—it needs a coordinated and effective approach to preserving and transferring this vital resource.

Countries with nuclear power programs have a significant responsibility for ensuring that **advanced nuclear science and engineering education and research programs** are in place. A coordinated approach to education, training, and outreach is necessary to demonstrate commitment to this important industrial sector and to assure that the new generation of nuclear workers is in place for all the ambition plans for new projects, constructions, and operations. Without **skilled and experienced workers**, there is a risk that nuclear knowledge creation will falter, placing unnecessary reliance on historical information to support new nuclear technology initiatives. For countries such as Vietnam, Thailand, and Malaysia, the insufficient local capacity makes it very difficult to maintain and develop their national regulatory frameworks and streamline resources toward nuclear power construction. This causes delays in the decision making about constructing or not, as well as delays in actual construction projects. Besides that, the reactor technologies originate usually from other vendor countries, such as Russia, Korea, or China. The challenge of being a knowledgeable customer, as well as consistent building up of the local expertise and decision-making capacity can be addressed primarily through collaboration in education, in regulating, and in dedicated training initiatives with the nuclear power countries. Establishing the sufficient framework is the main trend for nuclear knowledge management for new-comer countries. As nuclear power projects are often associated with substantial upfront investments, the countries should and do realize that more human resources are needed in non-power nuclear applications than in the power sector. Nuclear techniques are practiced in medicine, agriculture, water desalination, livestock, and much more. While the needs and requirements for nuclear technology may vary between those countries with **nuclear power installations** and those with only **non-power applications**, the need for nuclear knowledge management is essentially the same.

At the same time, Japan will need special efforts to preserve its wide nuclear expertise in the times of a temporary stagnation in the nuclear industry, as a consequence of the Fukushima Daiichi accident following the Great East Japan earthquake. Globally, Japan is not alone in a situation where the projections about the national nuclear industry lay between stagnation and phasing out. Germany took a decision on a gradual phasing out of nuclear power from its national energy mix. The main knowledge management-related difference between Germany and Japan is that the major part of German workforce is aging and will approach retirement around the time of phase out; however, Japan managed to assure a very balanced age distribution in its nuclear power. For continuous operation of nuclear facilities, this can be counted to a major knowledge management achievement. However, under current circumstances of uncertainty the nuclear knowledge management initiatives are integrated into other activities and are focusing on several tendencies which could help to preserve its wide nuclear expertise in the times of a temporary stagnation in the nuclear industry:

– Assuring projects for design, construction, and R&D organizations as sub-contractors in a supply chain for other projects in or led by other countries —this allows retention of design and construction-related expertise of people, technological upgrade, and keeping up-to-date with the rapidly developing standards for technology and management systems and the overall preservation of the national knowledge base.

- Expanding the experience in decommissioning based on the Plant Life Information Models and potential marketing of this expertise on the global market. Plant Life Information Modeling was critical for disassembling and decommissioning of the Fukushima Daiichi units. Exploring the technology-related aspect of nuclear knowledge management and its application throughout the life cycle of nuclear installation will be relevant for Japan for the next years.
- Informing the public, influencing public acceptance, and regaining trust from the direct customers of nuclear power are the tasks on the level of 'macro'—knowledge management, which has to be continuously conducted on national, regional, and prefecture level.
- Structured and consistent preservation of records and codified experience from the accident and from operations can be analyzed and considered in regulations and operations.

8.6.1 Looking to the Future

Nuclear Knowledge Management issues and priorities, especially related to the management of nuclear power plant (NPP) design knowledge organization of a knowledge management system or interfaces among knowledge organizations, are often unique for particular circumstances of individual country. They differ from organization to organization, from a regulatory authority to a technical support organization, EPC contractor, or NPP operating organization.

Nuclear Knowledge Management may focus on knowledge creation, identification, sharing, transfer, protection, validation, storage, dissemination, and preservation or utilization or tune all of these elements together. Nuclear knowledge management practices may enhance and support traditional business activities and can be initiated by organizational units which supervise human resource management, training, planning, operations, maintenance, quality assurance, and much more.

The application of **nuclear technology** is a fact of life in the twenty-first century. While its benefits are numerous, most applications involve working with hazardous material, so safety is paramount if this technology is to be utilized to the benefit of all. And **safety culture** depends on effective knowledge management.

The public reaction to Fukushima has shown us—yet again—that society's knowledge about nuclear energy and radiation is very weak. Since people do not yet understand it, the world's reaction was predictable. Public outcry affects government policy, which decides the future of nuclear power. This has been particularly seen in Asian region. Therefore, if nuclear power is to remain an energy option in future, it is not sufficient just to assure knowledge and skills needed to operate nuclear facilities. Nuclear knowledge managers need to find a way to address a broader group of people and inform and enlighten a receptive general public. Therein lays a huge challenge.

Our existing knowledge of the technology and its application must be preserved and, indeed, improved, if it is to play a central role in a global sustainable future. For doing this, nuclear knowledge has to be treated as a resource, intellectual, and knowledge-based assets of nuclear organizations should be handled consciously. Continued development of the technology for peaceful exploitation has to be based on the best knowledge available. It is of particular importance for the Asian region, as the main growth of nuclear power installed capacity is projected for countries here.

References

International Atomic Energy Agency. (1991). In *Safety Culture*. Safety Series Report No. 75, Vienna: IAEA.

International Atomic Energy Agency. (2006). *Knowledge management for nuclear industry operating organizations*. IAEA-TECDOC-1510.

International Atomic Energy Agency. (2007). *Milestones in the development of a national infrastructure for nuclear power*. No, NG-G-3.1, Vienna.

International Atomic Energy Agency. (2012a). *Knowledge management for nuclear research and development organizations*. IAEA-TECDOC-1675.

International Atomic Energy Agency. (2012b). *Managing nuclear knowledge—A pocket guide*. IAEA

International Atomic Energy Agency. (2016a). Nuclear accident knowledge taxonomy. In *IAEA nuclear energy series*, No. NG-T-6.8, Vienna: IAEA.

International Atomic Energy Agency. (2016b). *Leadership and Management for Safety*. Series No. GS-R Part 2, Vienna.

Karseka, T., & Yanev, Y. (2013). Knowledge Management as an Approach to Strengthen Safety Culture in Nuclear Organizations. *International Journal for Nuclear Power*, 58(4), April.

Schneider, M., & Froggatt, A., et al. (2016). *The world nuclear industry*, Status Report.

Part IV
Legal Issues

Chapter 9
Liability and Regulatory Aspects of Nuclear Energy Promotion in South Asia

M.P. Ram Mohan

9.1 Introduction

The thrust of movement towards nuclear energy is very much evident in the developing Asian region, the contemporary driver of atomic progress. Although there had been significant declines in nuclear projection across the world, the Middle East and non-OECD Asia have recorded substantial expansion. From 2000 to 2014, the non-OECD Asian region, which is rapidly developing, accounted for 55 of the 92 construction starts and 30 of the 53 new reactors that have been connected to the grid (IAEA 2014a). There is a visible shift in nuclear growth from North America and Europe to developing nations in Asia. The region is poised to be the driver of nuclear energy infrastructure in the coming decades. The Deputy Director General at the IAEA, Mikhail Chudakov, has stated that "Asia is one of the regions where nuclear is high on the agenda of a number of Member States and may be one of the drivers for global nuclear power deployment in the coming years" (Kollar 2016).

To a large extent, it was expected that the cataclysmic effects at Fukushima would deactivate Asia's nuclear quest. However, the developments in the region have proved otherwise (Basrur et al. 2015). There is still a continuing nuclear quest in Asia. Although the biggest plan for nuclear energy deployment is currently being carried out by China, the scope of this chapter does not cover that. The present study delves specifically into the South Asian region consisting of Afghanistan, Bangladesh, Bhutan, India, Maldives, Nepal, Pakistan and Sri Lanka. By signing an inter-governmental agreement, the countries in this region have formally established a geopolitical union under the umbrella term of the South Asian Association for Regional Cooperation (SAARC). Among the countries in this group, India and

M.P. Ram Mohan (✉)
Indian Institute of Management, Ahmedabad, India
e-mail: mprmohan@iima.ac.in

N. Janardhanan et al. (eds.), *Resurgence of Nuclear Power*,
https://doi.org/10.1007/978-981-10-5029-9_9

Pakistan already operate nuclear reactors for energy use with more under construction and visions of many in the pipeline. Bangladesh is a newcomer in the nuclear energy sphere and has commenced construction of its first nuclear power plant. Sri Lanka and Nepal, comparatively smaller nations, have also demonstrated their willingness to exploit nuclear energy by issuing national policies in this respect. The other three nations—Afghanistan, Bhutan and Maldives—have no long-term intention of operating nuclear power plants.

What makes the South Asian region unique in comparison to blocs in Europe is the fact that it is much more densely populated. The uniqueness of South Asia could further be understood in the context of nuclear non-proliferation. The Nuclear Non-Proliferation Treaty (NPT) is one of the most widely ratified global disarmament Convention in existence today. Although the NPT enjoys strong international support, there are five countries that have stayed outside the regime and two of them are in South Asia—India and Pakistan.[1] It is because of this reason that SAARC countries have not been able to establish a Nuclear-Weapons-Free Zone (NWFZ) in the region.[2] India and Pakistan continue to enhance nuclear military capabilities while also promoting peaceful applications of nuclear technology. India re-established its nuclear trade after receiving a waiver from the Nuclear Suppliers Group (NSG) in 2008. Consequently, it became the only State to have the capacity to legally collaborate internationally by staying outside the NPT regime.

These factors make South Asia a unique area for nuclear progress. By means of this chapter, the authors undertake to describe the nuclear scenario in South Asia. The paper would *firstly* report the kind of civil nuclear programmes that the countries have engaged in. *Secondly*, it would describe the international nuclear instruments that apply to them and check their adherence among the South Asian nations. And *thirdly*, it would detail the liability structures and regulatory set-ups that exist at the municipal levels. Through the course of this chapter, the authors bring to light the distinctive nature of nuclear arrangements in South Asia and the absence of a uniformly applicable framework in this respect.

9.2 South Asia Nuclear Programmes

This section highlights the background and the current status of nuclear energy projects in the South Asian region. Emphasis is laid upon India, Pakistan, Bangladesh, Nepal and Sri Lanka as these countries are either actively pursuing fission energy or have plans to introduce it in the future. Other countries like

[1]The other non-signatories are Israel, South Sudan and Democratic People's Republic of Korea (which withdrew from the treaty in 2003).

[2]A Nuclear-Weapons-Free Zone (NWFZ) is a specified region in which countries commit themselves not to manufacture, acquire, test or possess nuclear weapons. Article VII of the NPT permits states to specify and establish such zones. Currently, there are five NWFZs in the world.

Afghanistan, Bhutan and Maldives do not have a discernible intention of engaging in nuclear energy production and due of this reason, fall outside the scope of this section.

9.2.1 India

India's nuclear policies date back to the 1940s and envisage a long-term nuclear programme divided into three stages sequentially linked by fuel (Chidambaram 2006). India inaugurated its first nuclear reactor in 1969, and at present, there are 21 atomic power reactors in the country. These reactors constitute a total installed capacity of 5780 MW and operational experience of more than 418 years.[3] This has also had an effect on the share of nuclear power in the country's electricity mix. It climbed from 2.9% in 2004–05 to 3.6% in 2014–15, and the long-term goal is to strike 25% by 2050. In pursuance of that objective, the government expects to reach an installed nuclear capacity target of 13,480 MW by 2022 (Lok Sabha 2016a).

The latest Unit to be commissioned in India was a 1000 MW pressurized water reactor (PWR) built with Russian cooperation at Kudankulam, Tamil Nadu. This was the second Unit at the site. Public protests at Kudankulam NPP and questions of safety had earlier caused took the case beyond the coastal city and landed before the Supreme Court of India. After extensive hearing of parties, Supreme Court concluded the project is safe and the observed that the commissioning of the project should be seen as promotion of public welfare (more on the case discussion see, Ram Mohan and Shandilya 2014). The Indian government has now engaged Russian cooperation for two more Units (3 and 4) at Kudankulam. Similar expansion plans are also underway at existing nuclear power projects at Kakrapar in Gujarat and at Rawatbhata in Rajasthan. India's projected nuclear plan is ambitious and is detailed by means of Table 9.1.

The Jaitapur Nuclear Power Project (JNPP), slated to be the biggest civilian nuclear park in the world, is to be executed in collaboration with France. It would house six evolutionary pressurized reactor (EPR) units of 1650 MW each. In January 2016, France and India released a joint statement in which implementation of the project was aimed to begin in early 2017 (Joint Statement 2016). However, there exists disagreement between the French and Indian counterparts over two key elements: *firstly*, the cost of the reactors and *secondly*, the cost of electricity produced from the plant (Sasi 2016). The NPCIL is now looking at localization of components to capture further rises in cost.

Mithi Virdi in Gujarat was another site that was identified by the government for nuclear new-build. This location would have seen six reactors of Westinghouse AP1000 design of 1000 MW each. However, the land acquisition process at this

---

[3]This makes India 7th among 31 countries with nuclear power in terms of number of reactors and 13th in terms of electricity generation.

Table 9.1 Projected nuclear power projects in India

Indigenous reactors		Reactors with foreign cooperation	
Site and location	Capacity (MW)	Site and location	Capacity (MW)
Gorakhpur, Haryana	2 × 700[a]	Kudankulam, Tamil Nadu	2 × 1000[b]
Chutka, Madhya Pradesh	2 × 700	Jaitapur, Maharashtra	6 × 1650
Mahi Banswara, Rajasthan	4 × 700	Kovvada, Andhra Pradesh	6 × 1000[c]
Kaiga, Karnataka	2 × 700	Chhaya Mithi Virdi, Gujarat	6 × 1000[c]
Bhimpur, Madhya Pradesh	4 × 700	Haripur, West Bengal	6 × 1000[c]
Kalpakkam, Tamil Nadu	2 × 500		

Source Lok Sabha (2016b)
[a]In addition to GHAVP 1&2
[b]In addition to KKNPP 1–4
[c]Nominal Capacity

site did not go through which pushed the central government to look for an alternative site (Chaudhury 2016). This new site is Kovvada in Andhra Pradesh where Westinghouse will now deploy its nuclear reactors. The government has retained Mithi Virdi as a potential site for future plans.

India is still in the first stage of its civilian nuclear road map but is making veritable progress towards entering the second stage, that is of FBRs. State-owned operator, BHAVINI, has completed construction of a 500 MW Prototype FBR at Kalpakkam in Tamil Nadu that is at an advanced stage of commissioning (Lok Sabha 2016c). Being one its kind in the country, the reactor construction has raised challenges not only for the operator and suppliers but also for the regulator as it is technologically more complex than the reactors currently in operation in the country.

9.2.2 Pakistan

Pakistan's nuclear programme was initiated in 1955 when the Atomic Energy Committee was established. The country's policy was aimed at using nuclear technology in the field of energy, agriculture, health and industry (Mahmood 1995). At that time, Pakistan was extremely deficient in terms of sources of energy and quality of resources. It was importing a lot of oil from Gulf countries which resulted in large outflows of foreign exchange reserves. Pakistan acquired its first commercially operable reactor in the mid-1960s from Canada—a 137 MW PHWR which began commercial operation in 1971. Under the agreement, Canada was to provide both natural Uranium and heavy water in addition to technical and training assistance (Mahmood 1995). Currently, it has two more reactors in operation: Units—1&2 (300 MW each) at Chashma Nuclear Power Complex at Kudian, both of which were built by China at the time when it joined the Nuclear Suppliers Group (NSG) in 2004. Once it entered the supplier's cartel, it was expected that China would not deal with Pakistan again as it had not signed the NPT. However, in 2010, the China National Nuclear Cooperation announced that it would export technology for two new

Table 9.2 Status of nuclear power generation in Pakistan

Name and location	Capacity (MW)	Status
CHASNUPP-1, Kundian	325	Operational
CHASNUPP-2, Kundian	325	Operational
KANUPP, Karachi	137	Operational
CHASNUPP-3, Kundian	340	Under construction
CHASNUPP-4, Kundian	340	Under construction
K-2, Karachi	1100	Under construction

Source PRIS (2016)

reactors, Chashma-3 and Chashma-4 (315 MW each) because it argued that these projects were already grandfathered in under previous agreements rather than being fresh proposals (Parameswaran 2015). In particular, Beijing is helping Pakistan build these reactors in the background of the Obama administration supporting India in its nuclear progress. Pakistan moots that it has a right to the civil nuclear cooperation arrangement that the USA has with India. It has categorically stated that the country's energy needs were acute, power-generating reactors met international standards, and all the nuclear facilities were under IAEA safeguards and asserted a right to have access to civil nuclear technology (Iqbal 2015) (Table 9.2).

Pakistan's current installed nuclear generation capacity is 787 MW which contributed to 3.17% in the total nominal power generation capacity in the country (NEPRA 2015). The State has pushed for nuclear power generation by arguing that it promotes economic development and reduces reliance on foreign sources of energy. It expects that the development of nuclear in addition with coal projects would meet the necessary base load requirements in the overall demand profile of the country in the long run (Stephenson and Tynan 2009). Pakistan's Foreign Secretary, Aizaz Ahmad Chaudhary, has claimed that the country plans to generate 162 GW of electricity by 2030 out of which nuclear is expected to contribute 8.8 GW (Iqbal 2015). According to the country's post-Fukushima policy, there would be no change in nuclear power deployment plans up to 2030 (Mustafa 2012). In November 2013, it was further announced that nuclear energy will add 40 GW to the national grid by the year 2050 at an affordable cost (Kazmi 2014). This data is evidence of the country's increasing commitment to nuclear energy.

9.2.3 *Bangladesh*

Bangladesh is a nuclear newcomer and has been very careful in embracing nuclear technology. The country approved its Nuclear Power Action Plan in 2000 and identified Rooppur as the site for its first nuclear power plant. In 2010, the National Parliament in Bangladesh passed a resolution stating "In order to overcome the increasing power crisis in the country, nuclear power plant shall be established immediately" (Asaduzzaman and Akbar 2015). This is significant because it

Table 9.3 Projection of nuclear power generation in Bangladesh

Year	Site and Unit	Capacity (MW)
By 2020	Rooppur Unit-1	1000
	Rooppur Unit-2	1000
By 2025	Rooppur Unit-3	1000
By 2030	Rooppur Unit-4	1000

Source Bangladesh (2011)

exhibits public support for nuclear power as parliamentarians are representatives of the people in a democracy. It also demonstrates that the option of nuclear energy is sought to correct the power situation in the country. Later in 2015, the country signed a $12.65 billion deal with Russia to set up two NPPs at Rooppur, each of 1200 MW capacity. This is the country's biggest power project. The first reactor is expected to begin operating by 2022 and the second by 2023 (Ouadir 2015).

Bangladesh has laid great emphasis upon nuclear power and that is evident from the policies it has issued. For instance, in August 2016, the government passed an order exempting imported goods, machinery and parts for the Rooppur NPP from levy of tax (Rasel 2016). This exemption would predominantly benefit Atomstroyexport, the Russian contractor which is building the reactor. Bangladesh expects to derive 2 GW nuclear capacity by 2021 and scale it up to 4 GW by 2030 (Bangladesh 2011). For this purpose, it has planned an expansion of the existing site at Rooppur and has not identified any new sites (Table 9.3).

The country's vision about nuclear power is exemplified in the "Perspective Plan of Bangladesh (2010–2021)" whereby it seeks to derive 10% of its energy needs from nuclear power by 2021 (Planning Commission 2012). Bangladesh is currently in Stage II of the IAEA's Milestones Approach, which means it has completed the bidding process and has initiated construction on its site.[4] It has also kept a good track record of implementing the IAEA recommendations made during review missions (IAEA 2013).

However, the latest IAEA Integrated Nuclear Infrastructure Review (INIR) mission in June 2016 identified implementation of some recommendations that require further improvement such as strengthening the national project plan further to reflect the actual status of the programme and its future challenges. Other suggestions advanced by the team included finalization of a national and institutional human resource plan, issuance of a national communication strategy for stakeholder involvement and public information and formal approval of policies for the management of low- and medium-level radioactive waste (Dyck 2016). As Bangladesh moves ahead in its nuclear programme, it would encounter many challenges of such nature. But the positive side is that it is part of the NPT regime and that would ensure that it continually receives full support from the IAEA and other countries to address those problems in the coming times.

[4]The IAEA Milestones Approach was developed to assist countries that are considering or planning their first nuclear power plant.

9.2.4 Nepal

As of date, Nepal does not operate a nuclear reactor but does have a particular policy on exploiting nuclear energy. This policy is called the "National Nuclear Policy" and was issued in 2007. It was after releasing this strategy that Nepal became a member to the IAEA in 2008. The country maintains that it is not deterred from exploring peaceful applications of nuclear fission which go beyond the health sector. Nepal envisions to join the ranks of countries in its neighbourhood and become a producer of nuclear energy in the future. This intention is not primarily for the purpose of energy security but for larger interests such as economic and social prosperity, human health and world peace (Nepal 2007).

According to the policy document, there shall be a "Nuclear Steering Committee", within the aegis of the Ministry of Environment, Science and Technology, that will provide the necessary impetus in terms of directives and guidance regarding the implementation of the Policy for long-term basis (Nepal 2007). For development of a national nuclear law, the country intends to establish a Nuclear Technology Centre which will derive drafting assistance from the IAEA (MoST 2011). The law is slated to set the base for further concrete decisions for initiating an atomic energy programme. In 2016, Nepal signed a Country Programme Framework (CPF) with the IAEA for the period 2016–2021. A CPF is the frame of reference for the medium-term planning of technical cooperation between a Member State and the IAEA and identifies priority areas where the transfer of nuclear technology and technical cooperation resources will be directed to support national development goals (IAEA 2016). But Nepal's CPF did not include nuclear energy as a priority area. Evidently, the country is not concerned about production of atomic energy in the short term. Nepal intends to establish the basic level of infrastructure in terms of regulation, safety and security before embarking upon the more challenging activities of energy production which it has reserved for the future.

Nepal is surrounded by countries with active nuclear power programmes such as China, India, Pakistan and Bangladesh. But interestingly, it remains a non-party to all the major nuclear conventions covering a wide range of subjects, inter alia, safety, accident and liability. It is expected that an increased engagement with the IAEA coupled with its long-term plan of introducing nuclear power generation will make certain that Nepal joins the regimes that are necessary in the light of its goals.

9.2.5 Sri Lanka

Sri Lanka is the most recent participant in the nuclear race in South Asia. The country does not operate a nuclear power plant but does benefit from non-power uses of atomic energy for development purposes. Notwithstanding the foregoing, it does look at atomic energy with a positive perspective and intends to exploit it in the future. This is in the light of increasing growth and power demand in the island nation. The country has signed two CPFs with the IAEA in successive periods: 2009–2013 and 2014–2018. The 2009 CPF identified areas for near-term and medium-term programme, but it was specifically mentioned that contemplating nuclear power for electricity generation was the "long-term plan" for the country (IAEA 2009a). The 2014 CPF subsequently identified "sustainable energy development" as a priority area for its medium-term planning strategy. In 2012, the government of Sri Lanka appointed a steering committee to carry out a "Prefeasibility Study on the Use of Nuclear Power for Electricity Generation" (Sri Lanka 2014). This committee is expected to become the precursor of the Nuclear Energy Project Implementation Organization (NEPIO). The role of NEPIO in a State is to prepare for consideration or exploitation of nuclear energy (IAEA 2009b). In this capacity, the designated NEPIO not only leads the effort but also manages and coordinates the implementation of a nuclear power programme with other agencies in the country. The prefeasibility study would contemplate the option of using nuclear energy as a viable option beyond 2020. Subsequently in February 2015, the country signed an inter-governmental agreement with India on the subject of cooperation in the peaceful uses of nuclear energy. This was important for Sri Lanka as it was the first nuclear partnership for the country. The agreement would facilitate cooperation in transfer and exchange of knowledge and expertise, sharing of resources, capacity building and training of personnel in peaceful uses of nuclear energy including use of radioisotopes, nuclear safety, radiation safety, nuclear security, radioactive waste management and nuclear and radiological disaster mitigation and environmental protection (Agreement 2016). The agreement with India was followed by a similar nuclear cooperation agreement with Pakistan in April 2015.

9.3 International Nuclear Law in South Asia

The preceding section described the status of nuclear programmes in key countries of the South Asian region. The current section would outline the extent to which international nuclear law is applicable in the region and how the countries in this neighbourhood have helped to shape it.

9.3.1 Nuclear Non-proliferation

The Nuclear Non-Proliferation Treaty (*hereinafter*, NPT) is a landmark international treaty that aims to prevent the spread of nuclear weapons and weapons technology, foster the peaceful uses of nuclear energy and to further the goal of achieving nuclear disarmament.[5] It is the only international instrument that represents a binding commitment by the nuclear weapon states (countries that detonated a nuclear weapon before 1968) to the goal of disarmament.[6] There is no rigid timeline specified for realizing this objective, and in the Treaty's own language, it should be achieved "at the earliest possible date".

By means of their signature to the NPT, non-nuclear weapon States (non-NWS) were bound by the "never to acquire nuclear weapons" commitment. India refrained from agreeing to NPT terms and executed its peaceful nuclear explosion—Smiling Buddha—in 1974. In response to India's action, the nuclear countries formed a suppliers' cartel known as the Nuclear Suppliers Group (NSG). By issuing guidelines for nuclear and nuclear-related exports, NSG contributes to the non-proliferation of nuclear weapons thereby complimenting the work of the inter-governmental IAEA. In the South Asian region, two countries have persistently refused to join the NPT regime and those are India and Pakistan. The two countries regard the NPT as a failure and call it a discriminatory instrument owing to Treaty distinction between nuclear and non-nuclear weapon States.

India and Pakistan remain non-participants to two key nuclear disarmament treaties: NPT and CTBT (Comprehensive Test Ban Treaty). While NPT's objective is to "prevent the spread of nuclear weapons and weapons technology", CTBT aims to "prohibit any nuclear weapon test explosion or any other nuclear explosion". The international political lobby has made it easier for countries committed to disarmament to trade in nuclear materials and technology. However, in 2008, India managed to secure a waiver from NSG which recognized the country's long horizontal non-proliferation record (IAEA 2008). On top of that, India is the first and the only country for which NSG waived its guidelines. Now the country is looking to establish its position further by securing a membership to the NSG, a group where all the members are NPT parties as well. Among strong opposition, India's membership bid is supported by major countries including the USA, the UK, France, Germany, Russia, Japan, Canada, South Africa and Mexico.

In contrast to India and Pakistan, other South Asian countries have either ratified or acceded to the NPT: Afghanistan (1970), Bangladesh (1979), Bhutan (1985), Maldives (1970), Nepal (1970) and Sri Lanka (1979). They believe that the NPT is the cornerstone of non-proliferation which could be implemented only when all the countries commit themselves to complete disarmament (RevCon 2015). Bangladesh has declared that it supports the effective implementation of NPT but has called for universal adherence to the same. It believes that as long as nuclear weapons exist, so will the risk of their use. At the 2015 Review Conference of the NPT, Bangladesh (2015) urged the nuclear weapon states to fulfil, in good faith, their long overdue legal obligations under Article VI of the NPT to accomplish the complete elimination of their nuclear weapons. Article VI of the NPT mandates that

[5]These are generally referred to as the three pillars of the NPT.

[6]The NPT identified five nuclear weapon states (NWS) that also happen to be the five permanent members to the UN Security Council: the USA, the UK, France, Russia and China.

all states must pursue good faith negotiations for terminating the global nuclear arms race.[7] Furthermore, it names "nuclear disarmament under strict and effective international control" as the very specific objective of those negotiations. Simon (2004) notes that the obligation to think about and create a global treaty to complete nuclear disarmament still weighs on the nuclear weapons states. He asserts that they have a duty to negotiate in good faith and must create a subsequent and ultimate nuclear treaty that prohibits the creation of all new nuclear weapons and mandates the destruction of current stockpiles. Bangladesh has called for a comprehensive treaty that would ensure and guarantee the objective of a world free of nuclear weapons. It was also the first South Asian nation to ratify CTBT in 2000.

9.3.2 Safety and Emergency

International safety framework has largely emerged after the Chernobyl accident in 1986. Before that, States were not willing to entertain external interference in their nuclear programme. Post-Chernobyl, countries demonstrated tenacious momentum and adopted two instruments within a period of six months: The Convention on Early Notification of Nuclear Accident 1986 and the Convention on Assistance in the Case of a Nuclear Accident or Radiological Emergency 1986. The swiftness of international law in this case is remarkable and just goes on to show how committed the world was to espouse a framework that could help in responding to similar emergencies in the future.

Emergency situations demand that actions be taken by responsible organizations in a timely and effective manner to mitigate consequences on populations, infrastructure and environment, and support the return of affected areas to normal social and economic activity to the extent possible (NEA 2013). The international framework on nuclear emergency management rests entirely on the principle that a nuclear emergency anywhere is a nuclear emergency everywhere (NEA 2003). This is the reason for having a system of coordinated response that is guided by international agreements.

The Convention on Early Notification of Nuclear Accident (1986), via Article 1, establishes a notification system for nuclear accidents which have the potential for international transboundary release that could be of radiological safety significance for another State. Article 2 obliges States to report the accident's time, location, radiation releases and other data essential for assessing the situation. States must notify the affected State (directly or through the IAEA) and to the IAEA as wall. The Convention has 115 State Parties including India, Pakistan, Bangladesh, Sri Lanka and Afghanistan. India was disappointed with the outcome of the

[7]Treaty on the Non-Proliferation of Nuclear Weapons, Article VI: "Each of the Parties to the Treaty undertakes to pursue negotiations in good faith on effective measures relating to cessation of the nuclear arms race at an early date and to nuclear disarmament, and on a treaty on general and complete disarmament under strict and effective international control."

Convention and found it to be defective because it differentiated between nuclear and non-nuclear weapon States (India 1988). While it still ratified the treaty, the country made a declaration to express its dissatisfaction: "The Convention is defective as it does not contain a legal provision to make it mandatory on the nuclear weapon States to notify accidents involving nuclear weapons or weapon tests. The Government of India feels that the Convention should have provided for notification of nuclear accidents in any nuclear facility, vessel, aircraft, spacecraft, etc., used for peaceful or military purposes as well as nuclear weapons" (India 1988). Similar declaration was also made by numerous other Parties (including Sri Lanka) which compelled the P5 nations to declare that they would voluntarily notify the IAEA and any affected State in the event of accidents involving nuclear weapons and nuclear weapons tests (Declarations/Reservations 2016a).

Another issue that many countries comprehended was with respect to the dispute settlement procedure. Article 11(2) of the Convention allowed for submission of disputes arising out of the treaty to be submitted to arbitration or the International Court of Justice "at the request of any party to the dispute". In total, the Convention has received 50 State Party declarations, out of which 37 object to the same clause (including India, Pakistan and Sri Lanka). They submitted that any reference of a dispute for adjudication must be made by "all the concerned parties" to the dispute and not just "any party" to the dispute.

The Convention on Assistance in the Case of a Nuclear Accident or Radiological Emergency (1986), which was adopted at the same time as the foregoing Early Notification Convention, sets out an international framework for cooperation among State Parties and with the IAEA to facilitate prompt assistance and support in the event of nuclear accidents or radiological emergencies. It has 108 State Parties including India, Pakistan, Bangladesh, Sri Lanka and Afghanistan. Fifty-seven States have issued declarations with respect to this Convention stating that they are not bound by certain clauses. One of the causes of objection is the similarly drafted dispute settlement mechanism which India and Pakistan have refused to accept (Declarations/Reservations 2016b). Other South Asian countries—Bhutan, Maldives and Nepal—are not party to either of these two Emergency Conventions.

The aforementioned Conventions were an immediate response to the Chernobyl disaster. However, the post-Chernobyl review of nuclear energy production continued well into the 1990s. At an international conference in 1991, Member States to the IAEA called for a Convention to advance nuclear safety worldwide. This led to adoption of the Convention on Nuclear Safety (*hereinafter*, CNS) in 1994. The CNS is regarded as the most significant of the safety conventions in existence today. The 77 State Parties to the Convention include all the countries utilizing nuclear energy except Iran. In the South Asian context, Afghanistan, Bhutan, Maldives and Nepal have not signed the CNS. Before the CNS, nuclear safety was considered wholly a national legal matter. But the Chernobyl accident called into question the standards of safety to be implemented at nuclear power stations and the dangers of transboundary damage. There was a need for countries to commit to specific principles of nuclear safety in order to ensure that at least minimum parameters were applied across the globe. State Parties are legally bound to uphold a high level of nuclear safety, and the peer-review mechanism directed by the Convention helps to achieve this objective.

During the negotiation process leading up to the drawing up of CNS, there was lack of consensus among States as to whether the Convention should include in its scope the waste generated by nuclear reactors. This disagreement paved the way for Paragraph (ix) in the Preamble to the CNS (1994) which affirmed "the need to begin promptly the development of an international Convention on the safety of radioactive waste management as soon as the ongoing process to develop waste management safety fundamentals has resulted in broad international agreement". So the idea was to codify "soft law" in definite words in the language of international law.

In pursuance of this objective, the Joint Convention on the Safety of Spent Fuel Management and on the Safety of Radioactive Waste Management (*hereinafter*, Joint Convention) was adopted in 1997 and entered into force on 2001. The Convention combines two discrete subject matters, namely the safety of spent fuel management and the safety of radioactive waste management in a "joint" structure which constitutes its essential originality (Tonhauser and Jankowitsch-Prevor 1997). It was the first treaty to deal with these issues. Stoiber et al. (2003) states that the Joint Convention addresses three major difficulties. "The first is that radioactive waste will need to be managed safely well beyond the present generation. The second is that one State's radioactive waste may be another State's resource (in the Joint Convention, radioactive waste is defined as "radioactive material… for which no further use is foreseen by the Contracting Party…"). The third difficulty follows from the second one, in that some States reprocess nuclear fuel that has been irradiated in and then removed from nuclear reactors, and others do not: for those States that reprocess, the spent fuel is a resource, while for those that do not, it is radioactive waste". Tonhauser and Jankowitsch-Prevor (1997) note that it was in this context that India had proposed an amendment regarding the scope of the Joint Convention at the Diplomatic Conference in 1997. According to them, when India's proposal was adopted by votes, the scope of the Convention included safety of spent fuel management, in principle, but excluded the reprocessing activity unless the Contracting Party declared otherwise. This was different from what the UK had proposed—that the Joint Convention include the full range of spent fuel management and include reprocessing activity as well. India largely had two objections regarding the draft of the Joint Convention: *firstly*, the whole purpose of having a Convention on nuclear waste was to ensure human safety and the exclusion of military facilities from the treaty text diluted that objective; *secondly*, India considered spent fuel to be a resource and was not keen or enthusiastic about subscribing to a regime that dealt with both spent fuel and radioactive waste in the same document (BSCAL 1997). Moreover, the Indian delegation felt that the language of the treaty still maintained a distinction between nuclear and non-nuclear weapon States. As a result, India did not sign/ratify the Joint

Convention, and that status has not changed till now.[8] The other South Asian nations are also non-signatories to the Joint Convention. Recently, the Bangladesh Atomic Energy Commission has suggested that it is considering accessing the Joint Convention, but there are issues which it needs to address and is seeking counsel of the Ministry of Science and Technology in the country in this respect (Paul 2013).

9.3.3 *Liability*

Ensuring safety of a nuclear reactor or facility is immensely important. However, when things go wrong, the risk could lead to a transboundary event. Nuclear incidents cannot be said to be isolated events. Disaster in one State can lead to casualties in the neighbouring countries. In such cases, there would be a need to establish legal responsibility and determine liabilities. The IAEA believes that the goal of a nuclear liability Convention is to create treaty relations among States. "In this way, if there is an accident, it will be clear who is responsible and what framework applies" (Tonhauser and Ouevenco 2015). Currently, a State has an option to adhere to one or more of the following nuclear liability Conventions:

1. Paris Convention on Nuclear Third Party Liability (entered into force in 1968)
2. Vienna Convention on Civil Liability for Nuclear Damage (entered into force in 1977)
3. Convention on Supplementary Compensation (entered into force in 2015)

The aforementioned Conventions work on a set of principles which have since been accepted as globally applicable in international nuclear law. These principles are enumerated by Stoiber et al. (2003) as:

1. Strict liability or no-fault liability
2. Exclusive liability or legal channelling of liability on to the operator
3. Exonerations from liability[9]
4. Limitation of liability in amount
5. Limitation of liability in time
6. Financial security or congruence of liability and coverage

---

[8]In India, courts have often referred positively to treaties to which India remains a non-party. In *G. Sundarrajan v. Union of India*, (2013) 6 SCC 620, the Supreme Court had to determine whether the setting up of the Kudankulam Nuclear Power Project was against public policy. While declaring upon the priority of safety at nuclear installations, the Court drew extensively from the Joint Convention to provide context. According to the Bench "India is not a signatory to the same but the said Convention is worth referring to in order to understand and appreciate the worldwide concern for public safety" (paragraph 204).

[9]The operator will be exonerated from liability if it proves, for example, that the nuclear incident was directly due to an armed conflict, hostilities, civil war or insurrection, or that it resulted wholly or partly either from gross negligence of the victim or from an act or omission of the victim committed with intent to cause harm.

Table 9.4 Status of international nuclear law instruments in South Asia (as of September 2016)

South Asian countries	Nuclear power programme	NPT	CNS	Notification convention	Assistance convention	Vienna convention	CSC	Domestic law
Afghanistan	✗	✓	✗	✗	✓	✗	✗	✗
Bangladesh	✓	✓	✓	✓	✓	✗	✗	✓
Bhutan	✗	✓	✗	✓	✓	✗	✗	✗
India	✓	✗	✓	✓	✓	✗	✓	✓
Maldives	✗	✓	✗	✓	✓	✗	✗	✗
Nepal	✗	✓	✗	✓	✓	✗	✗	✗
Pakistan	✓	✗	✓	✓	✓	✗	✗	✓
Sri Lanka	Planning	✓	✓	✓	✓	✗	✗	✗

Source IAEA

7. Unity of jurisdiction
8. Non-discrimination

Notwithstanding the fact that the treaties espouse all the basic principles, the Paris Convention has 16 Contracting States which are mostly Western European, and the Vienna Convention has 40 Contracting States with Kazakhstan and Philippines as the only Asian members. The newest regime, the Convention on Supplementary Compensation, 1997 (*hereinafter*, CSC), is slated to fill this void. It was drawn up to provide the basis for a global nuclear liability regime that could attract broad adherence from countries with and without nuclear power plants (INLEX).

The CSC is open for membership by a country which is party to either the Paris Convention or Vienna Convention. If the State in question is not a Party to either of these regimes, then it must "declare" that its national law is in conformity with the provisions of the Annex to the CSC. But there exists a further proviso for a country to become an "Annex State". The additional condition is that it must also be a Contracting State to the CNS if it has an establishment on its territory that qualifies the definition of "nuclear installation" in the CNS.[10] Consequently, the Convention on Supplementary Compensation entered into force on 15 April 2015. As of September 2016, there are eight State Parties and 19 signatories to CSC. Japan and India are the only Asian countries to have ratified the Convention with the latter being the only South Asian nation to do so.

The non-existent enforceable international or regional legal instruments to deal with liability and compensation seem to expose the South Asian region to probable "Chernobyl" like situation. The legal vacuum is compounded with the fact that many countries within South Asia are choosing the nuclear energy path, thereby increasing the risk quotient (Ram Mohan et al. 2013) (Table 9.4).

[10]Recall that in the South Asian context, Afghanistan, Bhutan, Maldives and Nepal have not signed the CNS.

9.4 Nuclear Regulatory and Liability Regime in SAARC

9.4.1 Regulatory Regime

On the matter of regulation, international instruments are known to fix the basic requirements that each country should possess for nuclear regulation. IAEA believes that establishment of an independent and competent regulatory body is crucial to the long-term success of a nuclear energy programme (IAEA 2014b). This is also provided in international instruments. Article 8 of the CNS obliges a State to have a regulatory body that has adequate authority, competence and financial and human resources to fulfil the assigned responsibilities. It further states that "Each Contracting Party shall take the appropriate steps to ensure an effective separation between the functions of the regulatory body and those of any other body or organization concerned with the promotion or utilization of nuclear energy". Similar language is also used in Article 20 of the Joint Convention (1997). When it comes to responsibility for safety, it is the operator who bears the prime responsibility for the safety and security of the installation. On the other hand, the role of the regulatory body is to set standards and enforce them within the legal framework. The two Conventions referred to above do not stress upon a particular form of regulatory body—Commission form/Single Administrator or Directorate form—but they do stress upon the fact that it should be independent from bodies promoting or utilizing nuclear energy. It is important for both the regulator and the operator to pursue the common goal of safety and ensure risk-free operation of a nuclear installation.

9.4.2 India

Universally, safety has been the prime parameter in respect to nuclear energy generation since its very inception. In India, the primary responsibility for ensuring safety of NPPs rests with the operator (NPCIL), and the task of laying down safety criteria including rules and regulations is entrusted to the regulatory body, the Atomic Energy Regulatory Board (*hereinafter*, AERB) (Koley et al. 2006). The operation of NPPs is subject to meeting the AERB's rigorous safety requirements, which the government calls "adequate" and "favourably comparable" to other countries. A recent IRRS mission to India (IAEA 2015) also identified a number of good practices that, in their opinion, should be considered for implementation by other Member States (Ram Mohan and Kini 2015).

The AERB has issued Safety Codes for siting, design, quality assurance and operation of NPPs which are further supported by Safety Guides providing guidance on meeting the requirements of the Codes, while drafting of the standards due consideration is also given to the IAEA documents. By 2013, the AERB had published 141 documents relating to safety of the facilities in addition to the elaborate quality assurance mechanism to ensure the superiority of equipment supplied (Lok Sabha 2013). The norms call for mandatory adherence, and their non-observance is a penal offence as per the Atomic Energy Act, 1962, which punishes contravention of special safety provisions with imprisonment up to five years, or fine, or both. The AERB may also issue warnings or withdraw/suspend the license of a NPP which is a severe economic penalty having the potential to cripple the financial health of the stakeholders, which are the entities providing project finance (equity and debt).

Furthermore, India being a participant to the CNS is required to submit detailed national reports on the status of nuclear safety which are then peer-reviewed in an extensive manner. In 2012, the AERB became a party to the Multinational Design Evaluation Programme (MDEP) that develops innovative approaches to leverage the resource and knowledge of national regulatory authorities which undertake review of new reactor designs.

Indian NPPs are under continuous regulatory surveillance by the AERB. They undergo elaborate in-depth safety reviews during various consenting stages, viz. siting, construction, commissioning. Only after satisfactory review during project stage, the AERB issues operating license to a NPP, which is for a term of five years. Apart from this, extensive safety reviews are conducted for reactors which have been in service for over 25–30 years accounting for the age-related degradation.

Notwithstanding an all-encompassing safety framework highlighted in the foregoing paragraphs, a performance audit undertaken by the Comptroller and Auditor General of India (CAG) called for greater attention in monitoring regulated activities and recommended an independent legal structure as per international obligations (C&AG 2012). Appearing before the Public Accounts Committee (PAC) of the Parliament, AERB stated that the existing system of inspections for nuclear facilities was already rigorous and adequate. Further, they maintained that AERB was fully autonomous with regard to its regulatory decision-making and has not allowed its present legal status or organizational position, to hinder the discharge of the assigned responsibilities (PAC 2013).

In 2015, the IRRS mission to India had also recommended embedding the status of the nuclear regulatory authority in law (IAEA 2015). Consequently, the Nuclear Safety Regulatory Authority Bill was drafted to build greater credibility and inspire public trust (NSRA 2011).

The Nuclear Safety Regulatory Authority (*hereinafter*, NSRA) is conceived as a body with full powers to regulate nuclear safety in the country, reporting directly to the Parliament or a unit of Members of Parliament. This is expected to convert the functional independence of AERB to *de jure* independence. The proposal is currently awaiting completion of legislative process, but it fails to establish an airtight regulatory authority. Clause 5 of the Bill seeks to replace the AERB with the NSRA and would establish a Council of Nuclear Safety (CNS) which will be under the Prime Minister. The Council has the Chairman of the Atomic Energy Commission as one of its members. He is also the Secretary to the Department of Atomic Energy that controls NPCIL (the operator of NPPs in the country). However, this bill also has many critiques as to the structure as a full regulator. The problem seems to have arisen under Clause 10 where the Council is authorized to draw up Search Committees to select the Chairperson and other members to the NSRA. Thus, the Secretary of the DAE is involved in regulatory oversight and review of safety policy which is potentially an area of conflict of interest. Clause 14 suggests that even removal of members of the NSRA can be authorized by the central government without any judicial inquiry which further raises doubt about the independence of the new regulatory authority (NSRA 2011). The Parliamentary Standing Committee had also observed that the current Bill by and large seems to meet three out of the four core values viz. competence, independence, stringency and transparency, but it still lacks somewhat on the count of independence. The Committee found that there are certain clauses in the Bill viz. Clause 14(1) (removal of Chairperson and members of the authority), Clause 42 (directions by the central government to the authority) and Clause 48 (power of central government to supersede the authority), which may impinge on functional autonomy of the authority (standing committee 2012).

Further, the Bill seeks to amend the Right to Information Act, 2005, by including an additional clause (ca) under Section 8 of the Act. This would allow the NSRA to deny disclosure of confidential information, "sensitive information" and "commercially sensitive information of technology holders". Certain sections in the public believe that the government is further shielding the industry in this way. But inclusion of this new clause does not signify a major departure from general nuclear law. Countries are known to exercise restraint when dealing with information about physical protection of nuclear material, but it should be noted here that information that is relevant to safety of a plant must be disclosed and should not be shielded under any of the exemptions.

9.4.3 Pakistan

Pakistan Atomic Energy Commission (PAEC) was formed in 1965, and the regulatory activities were undertaken by divisions within the Commission. There did not exist a separate legal entity to deal with the regulatory functions. A nuclear safety and licensing division was established in PAEC HQ which functioned as the de facto regulatory body till it was upgraded to "Directorate of Nuclear Safety and Radiation Protection" (DNSRP) after the promulgation of Pakistan Nuclear Safety and Radiation Protection Ordinance 1984. When the country signed the CNS in 1994, it became obligatory upon Pakistan to establish a de jure entity to undertake the regulation of nuclear power and radiation use. Another requirement was to separate the regulatory function from the promotional aspect of nuclear energy. Consequently, the Pakistan Nuclear Regulatory Board (PNRB) was established in the same year as a provisional entity before the full legal separation was achieved by an Ordinance in 2001 which created the Pakistan Nuclear Regulatory Authority (*hereinafter*, PNRA). The new entity superseded both the PNRB and the

DNSRP. The 2001 Ordinance, the highest level document, describes the mandate, powers, functions and responsibilities of the PNRA assigned by the government.

The Pakistan Pakistan Nuclear Regulatory Authority is headed by a Chairman and constitutes two full-time and seven part-time members. The regulatory framework in Pakistan ensures nuclear and radiation safety by means of four types of documents in order of hierarchy: Ordinance, Regulations, Regulatory Guides and Codes and Standards. According to the World Nuclear Association (2016), the PNRA works closely with China's National Nuclear Safety Administration (NNSA) in respect to the Chashma reactors, and presumably also the Karachi coastal power project. In areas where the PNRA regulations are not available, latest regulations from the US Nuclear Regulatory Commission are applicable (PNRA 2016).

A recent IRRS mission to review the effectiveness of the PNRA concluded that Pakistan had a well-established regulatory and legal framework that was based on IAEA safety standards (IAEA 2014c). It noted that PNRA was an independent and competent regulatory body, empowered with the full scope of regulatory powers required by the IAEA standards. The mission identified some minor issues for which it gave certain recommendations as well.

9.4.4 Bangladesh

The entity responsible for promotion of nuclear activities in Bangladesh is the Bangladesh Atomic Energy Commission (*hereinafter*, BAEC). The Commission traces its origins to a Presidential order of 1973. In 1993, the government of Bangladesh promulgated the law on Nuclear Safety and Radiation Control (NSRC) in order to control and regulate the nuclear activities. In the absence of proper human resources and scientific expertise in the subject matter, the function of regulation was also handed over to the BAEC that already had that know-how instead of creating a new entity. This rendered BAEC with enormous powers as it was not only responsible for development of peaceful applications of nuclear technology but was also authorized to control the same. Subsequently, Bangladesh became a party to the CNS in 1995 that called for separation of the promotion functions from regulation.

In 2008, the first draft of the law was introduced and discussed at length. Consequently, in 2012, the Bangladesh Atomic Regulatory Authority Act was passed. This was done for the dual purpose of establishing an effective independent regulatory authority as well as for introducing nuclear law in the country. The legislation was formulated based on the IAEA Handbook of Nuclear Law and the existing NSRC Act of 1993. Inputs from several IAEA experts were also taken during formulating the draft of the Act. Provisions of the BAERA, 2012, inter alia, cover nuclear safety, security, safeguard of nuclear as well as radioactive materials (Chapter IV), transport and waste safety (Chapter V), emergency preparedness and remedial measures (Chapter VI), civil liability for nuclear damage in the event of an accident (Chapter VII) and inspection and enforcement (Chapter VIII).

The Bangladesh Atomic Energy Regulatory Authority (*hereinafter*, BAERA) was established under Section 4 of the Act on 12 February 2013 and was given the responsibility of regulating nuclear activities in the country. The BAERA is headed by a Chairman and consists of four other members who derive their power from the Act. The responsibilities and functions of the Authority are enumerated under Section 11 of the Act. The regulatory activities are carried out by the Authority by means of Acts/Ordinances, Rules/Regulations and Notifications, in that order of the hierarchy of documents. The statute also empowers the BAERA to form an Advisory Council to seek advice for the Authority on the scientific and regulatory aspects of nuclear safety and radiation protection (Section 16). This Council would again be headed by the Chairman of the Authority and consist of 13 members with six members from the scientific field and seven Joint Secretary level officers from various government departments. The BAERA is also going to start cooperation with India's regulatory authority, the AERB, under the ongoing "India Bangladesh Peaceful Nuclear Agreement-2016" (Choudhury 2016). Such exercises and training programmes would improve Bangladesh's human resource potential in nuclear regulatory activities.

9.5 Liability Regime in South Asia

The concerns within South Asia stem not only from the lack of a legal framework, but also from the conflicting liability frameworks under development (Ram Mohan 2015). This section would outline this very aspect of the existing as well as emerging nuclear law in South Asia.

9.5.1 India

Subsequent to the India specific IAEA Safeguards Agreement and NSG Wavier, the supplier countries demanded a set of liability rules based on the established principles of international nuclear liability law. Equally, a codified law was also seen as an important tool in alleviating public concerns. India, not being a signatory to either the Paris or the Vienna Convention, was keen on becoming a part of the CSC and joins the global nuclear liability regime. But in order to do that it had to ensure that it conceived a national law that was consistent with the provisions of that Convention. Therefore, in 2010, the Civil Liability for Nuclear Damages Act (*hereinafter*, CLND) received the Indian President's assent and became the law of the land. Civil liability law in India, prior to the enactment of the CLND, was based on judicially interpreted tort laws. So there existed no law to provide compensation to victims of a nuclear incident. Therefore, the new law was expected to provide an unequivocal and rule-based liability system.

Enactment of a nuclear liability legislation was sought to assist India in joining the CSC and also facilitate its collaboration with other countries. Though the government insists that the basic elements of the CLND are compliant with the CSC, which India signed in 2010, many supplier countries led by the USA believe otherwise. Section 4 of the Act channels liability for nuclear damage to the operator of the nuclear installation (based on the principle of no-fault liability). If there is a nuclear incident at a reactor, Section 6(2) states that the operator could be liable up to INR 1500 crores in that case. Once the operator of the nuclear installation dispense compensation to the victims (claims are to be filed before a Claims Commissioner and not a civil court), the law provides that it can seek recourse against the supplier in accordance with three situations under Section 17 of the Act. Grover (2016) states that the approach taken by CLND towards supplier's liability is different from that taken by other countries, and this has led to an intense debate within India and by the global community specializing in nuclear law. The specific deviance from the international framework, alleged by suppliers, both domestic and foreign, is Section 17(b), where the operator could exercise this right if the accident was the result of a patent or latent defect in equipment or substandard service. The supplier corporations conceded this clause as violative of international law. Moreover, the Act by means of Section 46 gave the idea that claims for compensation under other laws could be brought against operator and suppliers. This implied that the liability ceiling specified in the Act could be rendered a variable financially distressing the operator and suppliers even more. Therefore, following the passage of the Act, few potential suppliers requested India to adopt a domestic regime in line with established international practices while citing Sections 17 and 46 of the Act. Nuclear law experts, such as Pelzer (2011), also agreed with the fact the CLND was a "flawed legislation" and that India cannot, on the basis of the Act, "become party to any of the existing international nuclear liability treaties, because it does not meet their basic requirements."

However, the government has now cleared this ambiguity by stating that suppliers were never covered under Section 46 and the remedy is available only against the operator, thereby maintaining the globally accepted concept of channelization of nuclear liability. Furthermore, the government has also declared that Section 17 is "not a mandatory but an enabling provision". So unless the operator includes the right of recourse in its contract with the supplier, it cannot exercise the same in any event whatsoever (MEA 2015). Depending upon the contract, the operator and supplier can agree upon the recourse mechanism. Although the operator is required to furnish financial security covering its liability under the Act, the suppliers were concerned that there was no cover for their risks. The government, while acknowledging the complications, iterated numerously that it did not propose a change in the legislation anytime soon. However, in February 2015 (immediately after President Obama's visit to India in January), a press release in the form of frequently asked questions and answers (FAQ) from the Ministry of External Affairs provided much needed clarity in this respect (MEA 2015). Among many other explanations, the FAQ document cited that the India Nuclear Insurance Pool (INIP) had been established to allay the concerns of suppliers. The Pool was launched on 12 June 2015, to provide the necessary source of funds for the suppliers to undertake cover for their risk exposure in the event of invocation of the right of recourse provision (PIB 2016).

9.5.2 Pakistan

Pakistan's civil liability law is contained in the same Ordinance that established the country's regulatory body, the Pakistan Nuclear Regulatory Authority (PNRA). The Preamble to the Ordinance states that the legislation provides for "the extent of civil liability for nuclear damage resulting from any nuclear incident". So there exists no separately distinct legislation addressing the matter of civil liability. Moreover, Pakistan is neither a member of any international Convention on nuclear liability nor is it associated with any nuclear insurance pool. Consequently, the issue of civil nuclear liability in Pakistan is squarely covered by Sections 30–37 of the 2001 Ordinance.

Section 30 of the Ordinance states that "The Authority shall have the power to fix the extent of civil liability for an operator in case of various nuclear incidents". As a result, the statute does not affix an amount itself. The law further channels the liability absolutely to the operator in case of damage caused due to a nuclear incident. An operator in Pakistan is also required to undertake insurance or other financial guarantee adequate to cover its civil liability for nuclear damage. If the liability extends beyond any such cover, then it will be the responsibility of the government to ensure the payment of the remaining claim amount. Unlike the Indian law, the PNRA does not create a right of recourse for the operator. Section 35 of the Ordinance also covers instances of transboundary damage and suggests that if any nuclear incident in Pakistan causes damage in another State, the jurisdiction of claims under the Ordinance shall rest with the competent courts in Pakistan. Section 36 is the limitation clause that declares that a right to compensation shall stand extinguished if the claim is not made within 10 years of the nuclear incident and within three years from the date on which the person suffering nuclear damage had knowledge of the nuclear damage. It is further provided that the period shall in no case exceed ten years. However, to this extent, there is a departure from international nuclear liability principle. Radiation damage may be latent for a long time, and that is why international nuclear law specifies a minimum period of 30 years in the event of personal injury caused by nuclear damage. Ten years would be too short a period of time to detect radiation damage.

9.5.3 Bangladesh

The Bangladesh liability regime is similar in structure to Pakistan's arrangement. There is no separate legislation dealing specially with the topic of civil nuclear liability but is covered within the law that established the nuclear regulatory authority. Accordingly, Chapter VII of the Bangladesh Atomic Energy Regulatory Act of 2012 lays out the civil liability law applicable in the country. The statute legally channelizes the liability of nuclear damage due to the operator of the nuclear installation which caused the nuclear incident. The definition of "nuclear damage" is largely based upon the definition of the term laid down under the Indian liability law but is a slight improvement in terms of clarity. Section 45 of the Act caps the liability amount with respect to each nuclear incident at Bangladesh Taka equivalent of 300 million SDRs, which is in consonance with the minimum amount specified under the liability regimes of the 1997 Vienna Protocol and the CSC. The provisions allow for a claim to be brought in a civil court in Bangladesh for nuclear damage that is caused by a nuclear incident occurring within the territory or within the exclusive economic zone of Bangladesh. This is in compliance with the international principle of unity of jurisdiction where the courts of accident have exclusive jurisdiction in case of compensation claims. Furthermore, there does not exist a right of recourse in the BAERA legislation.

The interesting departure is, however, in the limitation clause. The BAERA declares that the right to compensation is exhausted if the claim is not brought before 50 years (in case of loss of life or personal injury) and 25 years (in case of any other nuclear damage) from the date of the nuclear incident. Such a right shall also stand extinguished 10 years from the date on which the victim had knowledge or ought reasonably to have had knowledge of the damage. This is a huge extension from what the legislations in India or Pakistan provide. For a nuclear newcomer to have such an open policy on liability is a novel development and is definitely worthy of praise.

9.6 Conclusion

South Asia is one of the few regions in the world wherein countries are in ambitious pursuit of the nuclear dream to meet the ever-increasing electricity demand in their territories. Even the Fukushima accident has had negligible effect on atomic energy projections in the region. Among the South Asian nations, India has the most ambitious nuclear energy programme followed by projections made by Pakistan. These are also the countries that have stayed out of the NPT and CTBT regime. Bangladesh, being the latest nuclear newcomer in the region, Sri Lanka is still in the decision phase, while countries including Afghanistan, Bhutan, Maldives and Nepal have no nuclear plans.

One of the most important parts of any nuclear programme is a strong competent and efficient regulatory system which is capable of taking decisions irrespective of any extraneous considerations. India and Pakistan have nuclear regulatory institutions that are working, and which has been recently audited by IAEA as well. Bangladesh is currently setting up its regulatory infrastructure. The nuclear power-producing countries in South Asia, while building the capabilities on learnings from many countries must head and remember what the Fukushima Nuclear Accident Independent Investigation Commission of Japan concluded: "The TEPCO Fukushima Nuclear Power Plant accident was the result of collusion between the government, the regulators and TEPCO, and the lack of governance by said parties. They effectively betrayed the nation's right to be safe from nuclear accidents".

On the question of liability, countries in South Asia are divided over which liability Convention to sign up to. India, Bangladesh and Pakistan have municipal nuclear liability laws. Out of these, India is the only country to be a member of an international nuclear liability regime, the CSC. Bangladesh has drafted its domestic law in consonance with the Vienna Convention but has not signed the same. Pakistan's nuclear liability law leaves out many key components to the executive and is not as airtight as the ones in preceding countries. The subject matter of liability has clearly emerged as a variable in the South Asian context, and there is a void that needs to be filled. This vacuum also puts a question mark on future nuclear new-build in the region. Therefore, it is important that countries become party to a nuclear liability regime to address the issues arising out of transboundary damage in the event of a nuclear incident. This becomes exceedingly significant as more and more countries in South Asia plan on exploiting nuclear energy in the coming years.

Acknowledgements I am thankful to Akshay Shandilya, LL.M student at TERI University for providing extensive research support towards completing the paper.

References

Agreement. (2016). Agreement between the Republic of India and the democratic socialist Republic of Sri Lanka on cooperation in the peaceful uses of nuclear energy. http://dae.nic.in/writereaddata/ncpw/IGA_srilanka_2015.pdf. Accessed September 6, 2016.

Asaduzzaman, K. M., & Akbar, M. S. (2015). National nuclear power programme of Bangladesh. *Technical Meeting on Topical Issues in the Development of Nuclear Power Infrastructure*. Vienna: IAEA. https://www.iaea.org/NuclearPower/Downloadable/Meetings/2014/2015-02-03-02-06/D1_S2_Bangladesh_Akbar.pdf. Accessed September 10, 2016.

Bangladesh. (2011). Power system master plan 2010. *Government of the People's Republic of Bangladesh*. Table 8–19.

Bangladesh. (2015). Statement by H.E. Mr. M. Shameem Ahsan. In *Review Conference of the Nuclear Non-Proliferation Treaty (NPT)*.

Basrur, R., Chang, Y., & Koh, S. L. C. (2015). Nuclear energy in Asia: The end of a Renaissance? In P. G. Harris & G. Lang (Eds.), *Routledge handbook of environment and society in Asia* (pp. 423–438). London: Routledge.

BSCAL. (1997). *India Opposes Classification of Spent Fuel as Nuclear Waste. Business Standard*. http://www.business-standard.com/article/specials/india-opposes-classification-of-spent-fuel-as-nuclear-waste-197090301080_1.html. Accessed September 8, 2016.

C&AG. (2012). Performance audit of the activities of atomic energy regulatory board for the year ended March 2012. *Report No. 9*. http://www.cag.gov.in/sites/default/files/audit_report_files/

Union_Performance_Atomic_Energy_Regulatory_Board_Union_Government_Atomic_Energy_Department_9_2012.pdf. Accessed September 11, 2016.
Chaudhury, D. R. (2016). US firm westinghouse likely to shift its nuclear plant site plan from Gujarat to Andhra Pradesh. *The Economic Times*. http://economictimes.indiatimes.com/news/politics-and-nation/us-firm-westinghouse-likely-to-shift-its-nuclear-plant-site-plan-from-gujarat-to-andhra-pradesh/articleshow/52578747.cms#. Accessed August 20, 2016.
Chidambaram, R. (2006). India and the international atomic energy agency. *Atoms for Peace: An International Journal*, *1*(2/3).
Choudhury, N. (2016). Preparing to license the RNPP site licence and construction licence—Nuclear regulatory status of BAERA considering IAEA 19 infrastructure issues [NG-G-3.1 (Rev. 1)]. Vienna: IAEA. https://www.iaea.org/NuclearPower/Downloadable/Meetings/2016/2016-02-02-02-05-NIDS/S4_1_Choudnury_Bangladesh.pdf. Accessed September 10, 2016.
CLND. (2010). The Civil Liability for Nuclear Damages Act. Act No. 38 of 2010.
Convention on Assistance in the Case of a Nuclear Accident or Radiological Emergency. (1986). INFCIRC/336.
Convention on Early Notification of Nuclear Accident. (1986). INFCIRC/335.
Convention on Nuclear Safety. (1994). INFCIRC/449.
Declarations/Reservations. (2016a). Convention on early notification of a nuclear accident. *IAEA*. https://www.iaea.org/Publications/Documents/Conventions/cenna_reserv.pdf. Accessed September 2, 2016.
Declarations/Reservations. (2016b). Convention on assistance in the case of a nuclear accident or radiological emergency. *IAEA*.
Dyck, E. (2016). IAEA reviews progress of Bangladesh's nuclear infrastructure development. *IAEA*. https://www.iaea.org/newscenter/news/iaea-reviews-progress-of-bangladeshs-nuclear-infrastructure-development. Accessed September 5, 2016.
Grover, R. B. (2016). Opening up of international civil nuclear cooperation with India and related developments. *Progress in Nuclear Energy*. doi:10.1016/j.pnucene.2016.09.016.
IAEA. (2008). Statement on civil nuclear cooperation with India. INFCIRC/734.
IAEA. (2009a). Sri Lanka signs CPF. https://www.iaea.org/technicalcooperation/Regions/Asia-and-the-Pacific/News/Archive/Sri_-Lanka-CPF.html. Accessed September 5, 2016.
IAEA. (2009b). Responsibilities and capabilities of a nuclear energy programme implementing organization. *IAEA Nuclear Energy Series No. NG-T-3.6*. Vienna: IAEA.
IAEA. (2013). Progress in building Bangladesh's nuclear infrastructure. https://www.iaea.org/NuclearPower/News/2013/2013-10-11-inig.html. Accessed August 4, 2016.
IAEA. (2014a). International status and prospects for nuclear power 2014. *Report by the Director General of the IAEA*. GOV/INF/2014/13-GC(58)/INF/6.
IAEA. (2014b). *Developing infrastructure for new nuclear programmes*. Vienna: IAEA.
IAEA. (2014c). IAEA mission concludes peer review of Pakistan's nuclear regulatory framework. https://www.iaea.org/newscenter/pressreleases/iaea-mission-concludes-peer-review-pakistans-nuclear-regulatory-framework. Accessed August 10, 2016.
IAEA. (2015). Report of the integrated regulatory review service (IRRS) mission to India. http://www.aerb.gov.in/AERBPortal/pages/English/t/documents/irrs.pdf. Accessed September 5, 2016.
IAEA. (2016). Nepal signs a country programme framework (CPF) for 2016–2021. https://www.iaea.org/technicalcooperation/Home/Highlights…/03022016_nepal.html. Accessed September 1, 2016.
India. (1988). Declaration of the republic of India made upon expressing consent to be bound by the convention on early notification of a nuclear accident.
INLEX. (2016). Civil liability for nuclear damage: Advantages and disadvantages of joining the international nuclear. *International Expert Group on Nuclear Liability*. https://ola.iaea.org/ola/treaties/documents/liability_regime.pdf. Accessed August 31, 2016.
Iqbal, A. (2015). Pakistan will not sign NPT, says foreign secy. *Dawn*. http://www.dawn.com/news/1185843. Accessed September 2, 2016.
Joint Convention on the Safety of Spent Fuel Management and the Safety of Radioactive Waste Management. (1997). INFCIRC/546.

Joint Statement. (2016). *India-France Joint Statement on the occasion of the State Visit of President Francois Hollande of the French Republic to India.*

Kazmi, Z. (2014). Pakistan's energy security. *The Express Tribune*. http://tribune.com.pk/story/655573/pakistans-energy-security. Accessed September 10, 2016.

Koley, J., Harikumar, S., Ashraf, S. A. H., Chande, S. K., & Sharma, S. K. (2006). Regulatory practices for nuclear power plants in India. *Nuclear Engineering and Design, 236*(7–8), 894–913.

Kollar, L. (2016). Asia's prospects for nuclear power highlighted at regional conference. *IAEA*. https://www.iaea.org/newscenter/news/asias-prospects-for-nuclear-power-highlighted-at-regional-conference. Accessed September 1, 2016.

Lok Sabha. (2013). Question No. 544. Monsoon Session. http://www.dae.nic.in/writereaddata/parl/mansoon2013/lsus544.pdf. Accessed September 12, 2016.

Lok Sabha. (2016a). Question No. 1078. Budget Session. http://www.dae.nic.in/writereaddata/parl/budget2016/lsus1078.pdf. Accessed September 12, 2016.

Lok Sabha. (2016b). Question No. 2784. Monsoon Session. http://dae.nic.in/writereaddata/parl/monsoon2016/lsus2784.pdf. Accessed September 12, 2016.

Lok Sabha. (2016c). Question No. 1762. Budget Session. http://www.dae.nic.in/writereaddata/parl/budget2016/lsus1762.pdf. Accessed September 12, 2016.

Mahmood, T. (1995). Nuclear non-proliferation treaty (NPT): Pakistan and India. *Pakistan Horizon, 48*(3), 81–100.

MEA. (2015). Frequently Asked Questions and Answers on Civil Liability for Nuclear Damage Act 2010 and Related Issues. *Ministry of External Affairs—India*. http://www.mea.gov.in/press-releases.htm?dtl/24766/Frequently_Asked_Questions_and_Answers_on_Civil_Liability_for_Nuclear_Damage_Act_2010_and_related_issues. Accessed August 2, 2016.

MoST. (2011). Report of the National Workshop on Application of Nuclear Technology in Nepal. *Jointly organized by the Ministry of Science and Technology, Nepal, and Pokhara University.*

Mustafa, G. (2012). Prospects of nuclear power in Pakistan. *INPRO Dialogue Forum on Global Nuclear Energy Sustainability Long-Term Prospects for Nuclear Energy in the Post Fukushima Era*. Seoul: IAEA. https://www.iaea.org/INPRO/5th_Dialogue_Forum/Tuesday,_28.08.2012/1110-1230(National_Perspective)/6._Ghulam_Mustafa_Pakistan_0828.pdf. Accessed September 12, 2016.

NEA. (2003). Case study 2—Nuclear accidents. In *Emerging Risks in the 21st Century: An Agenda for Action, OECD*, 203–204.

NEA. (2013). Strategic aspects of nuclear and radiological emergency management. *OECD*. https://www.oecd-nea.org/rp/inex/strategic-aspects.html. Accessed September 1, 2016.

Nepal. (2007). National Nuclear Policy (2064).

NEPRA. (2015). State of Industry Report.

NSRA. (2011). Nuclear Safety Regulatory Authority Bill.

PAC. (2013). Activities of atomic energy regulatory board. *Public Accounts Committee*. http://164.100.47.134/lsscommittee/Public%20Accounts/15_Public_Accounts_90.pdf. Accessed September 11, 2016.

Parameswaran, P. (2015). China confirms Pakistan nuclear projects. *The Diplomat*. http://thediplomat.com/2015/02/china-confirms-pakistan-nuclear-projects. Accessed September 11, 2016.

Paul, D. (2013). Current status of the radioactive waste management, issues to be overcome, and the way forward to access the JC. *Regional Workshop on the Joint Convention on the Safety of Spent Fuel Management and on the Safety of Radioactive Waste Management*. Seoul: IAEA. https://ansn.iaea.org/Common/Topics/OpenTopic.aspx?ID=13525. Accessed September 11, 2016.

Pelzer, N. (2011). The Indian Civil Liability for Nuclear Damage Act, 2010—Legislation with Flaws?. *atw—International Journal for Nuclear Power* 56(1): 1–9. http://www.kernenergie.de/kernenergie-wAssets/docs/fachzeitschrift-atw/2011/atw2011_01_pelzer_indian_civil_liability.pdf. Accessed September 8, 2016.

PIB. (2016). Launching of nuclear insurance pool. *Press Information Bureau*. http://pib.nic.in/newsite/PrintRelease.aspx?relid=137276. Accessed August 2, 2016.

Planning Commission. (2012). Perspective plan of Bangladesh 2010-202: Making vision 2021 a reality. *Government of the People's Republic of Bangladesh.*

PNRA. (2016). Regulatory framework. *Pakistan Nuclear Regulatory Authority*. http://www.pnra.org/r-framwork.html. Accessed September 5, 2016.

PRIS. (2016). Power reactor information system. *IAEA*.

Quadir, S. (2015). Bangladesh signs $12.65 billion deal with Russia for nuclear power plants. *Reuters*. http://uk.reuters.com/article/uk-bangladesh-russia-power-idUKKBN0U80TQ20151225. Accessed September 14, 2016.

Ram Mohan, M. P. (2015). *Nuclear energy and liability in South Asia: Institutions, legal frameworks and risk assessment within SAARC*. New Delhi: Springer.

Ram Mohan, M. P., & Kini, E. R. (2015). India's nuclear regulators have been audited. *Hindu Business Line*. http://www.thehindubusinessline.com/opinion/indias-nuclear-regulators-have-been-audited/article8061473.ece. Accessed September 7, 2016.

Ram Mohan, M. P., Raju, K. D., & Shiju, M. V. (2013). A nuclear liability framework for South Asia: Formation of South Asia association for regional cooperation (SAARC) nuclear risk community. *International Journal of Nuclear Law, 4*(1), 46–62.

Ram Mohan, M. P., & Shandilya, A. (2014). Nuclear energy and risk assessment by Indian Courts: Analysis of judicial intervention in the Kudankulam nuclear power project. *Journal of Risk Research, 18*(8), 1051–1069.

Rasel, A. R. (2016). Rooppur nuclear power plant gets tax exemption. *Dhaka Tribune*. http://www.dhakatribune.com/bangladesh/2016/09/12/rooppur-nuclear-power-plant-gets-tax-exemption. Accessed September 14, 2016.

RevCon. (2015). Joint statement on the humanitarian consequences of nuclear weapons. *2015 Review Conference of the Parties to the Treaty on the Non-Proliferation of Nuclear Weapons*.

Sasi, A. (2016). Jaitapur nuclear project: Renewed push, amid lingering concerns. *The Indian Express*. http://indianexpress.com/article/business/business-others/jaitapur-nuclear-project-renewed-push-amid-lingering-concerns/. Accessed September 12, 2016.

Simon, D. (2004). Article VI of the non-proliferation treaty is a pactum de contrahendo and has serious legal obligation by implication. *Journal of International Law & Policy, University of Pennsylvania, Vol 12,* 1–19.

Sri Lanka. (2014). Country specific framework 2014–2018. *Government of Sri Lanka*. http://aeb.gov.lk/web/attachments/article/299/SRL_CPF2014-17_Master_Document_FINAL.pdf. Accessed September 2, 2016.

Standing Committee. (2012) 221st report on the nuclear safety regulatory authority bill, 2011. *Department related Parliamentary Standing Committee on Science and Technology, Environmental and Forests*.

Stephenson, J., & Tynan, P. (2009). Is nuclear power Pakistan's best energy investment? Assessing Pakistan's electricity situation. In Henry D. Sokolski (Ed.), *Pakistan's Nuclear Future: Reining in the Risk* (pp. 103–130). U.S. Army War College: Strategic Studies Institute.

Stoiber, C., Baer, A., Pelzer, N., & Tonhauser, W. (2003). *Handbook on nuclear law*. Vienna: IAEA.

Tonhauser, W., & Jankowitsch-Prevor, O. (1997). Joint convention on the safety of spent fuel management and on the safety of radioactive waste management. *Nuclear Law Bulletin, 60,* 9–22.

Tonhauser, W., & Quevenco, R. (2015). Towards a better understanding of civil liability for nuclear damage: Learning the fine points of the international legal regime. *IAEA*. https://www.iaea.org/newscenter/news/towards-better-understanding-civil-liability-nuclear-damage-learning-fine-points-international-legal-regime. Accessed September 10, 2016.

World Nuclear Association. (2016). Nuclear power in Pakistan. *World Nuclear Association*. http://www.world-nuclear.org/information-library/country-profiles/countries-o-s/pakistan.aspx. Accessed September 8, 2016.

Author Biography

Dr. M.P Ram Mohan is an Associate Professor at Indian Institute of Management Ahmedabad. Over the last few years, Dr Ram Mohan's primary area of work has been in the area of nuclear energy and its interaction with policy and society; law and public policy, risk regulation and governance, and state/industrial liability and law.

Dr. Mohan is a nominated member of the "Nuclear energy for peaceful applications committee", Bureau of Indian Standards, Government of India, that formulates Indian standards for nuclear energy. He is the President of the Nuclear Law Association, India and International Union for Conservation of Nature (IUCN) has recognized him as the India Focal point of IUCN World Commission on Environmental Law and is a member of its Energy Law and Climate Change Specialist Group. Prior to joining IIM Ahmedabad, he was associated with TERI University (as Associate Professor); TERI, New Delhi (as Fellow) and National University of Juridical Sciences Kolkata (as Research Faculty). He started his career as an advocate at the Trivandrum District Court, Kerala and has authored "Nuclear Energy and Liability in South Asia: Institutions, Legal Frameworks and Risk Assessment within SAARC" published by Springer in 2015 (Dr. Ram Mohan can be contacted at: mprmohan@iima.ac.in).

Part V
Economics

Chapter 10
A Perspective on the Cost of Nuclear Energy

Vaibhav Chaturvedi, Priyadarshi R. Shukla and Karthik Ganesan

10.1 Introduction

India's energy demand has witnessed a significant growth in the recent past. Commercial primary energy demand, excluding biomass, has increased from 300 Mtoe in 2000 to 589 Mtoe in 2013 (IEA 2015), reflecting an average increase of 5.33% per annum. With electricity consumption in India at 1010 kWh/capita (Bhaskar 2015) in 2014–2015 being less than 15% of that of the developed economies (USA, Germany, Japan, etc.), significant growth in India's electricity generation capacity is required in the near and medium term to meet the growing demand.

Importance of nuclear energy for meeting India's energy challenges in the near as well as long term has been highlighted earlier in a number of studies (e.g. see Mallah and Bansal 2010; Mohapatra and Mohanakrishnan 2010; GoI 2011; Remme et al. 2011; IAEA 2012; Shukla and Chaturvedi 2012; Kour and Dar 2013; Shukla and Chaturvedi 2013). Inter-model comparison studies have shown that nuclear energy plays a significant role in India's long-term electricity production mix across energy models especially under climate policy scenarios (Clarke et al. 2012). The official government policy also has emphasized the importance of this energy source. This has been reflected in hectic efforts made by the Indian government to legislate the nuclear energy deal as well as in the increased pace of related

This chapter is based on an ongoing research the author(s) have been involved since 2014 with CEEW. The chapter includes updated information with in-depth analysis based on latest data on the cost of nuclear energy and other power generation technologies.

V. Chaturvedi (✉) · K. Ganesan
Council on Energy, Environment and Water (CEEW), New Delhi, India
e-mail: vaibhav.chaturvedi@ceew.in

P.R. Shukla
Indian Institute of Management Ahmedabad, Ahmedabad, India

N. Janardhanan et al. (eds.), *Resurgence of Nuclear Power*,
https://doi.org/10.1007/978-981-10-5029-9_10

negotiations with governments and businesses around the world for ensuring supply of nuclear technology and fuel.

While the efforts reflect that the government and numerous experts are optimistic about the future of nuclear energy in India, strong concerns remain in India and globally about the costs and risks of nuclear energy (Ramana 2009; Sokolski 2010; Abraham 2011; Chang et al. 2011; Hibbs 2012). Ramana (2012) in his recent book reflects these concerns and argues against making nuclear energy an important pillar of India's energy supply policy. The argument is largely based on the history of nuclear disasters even in the most technologically advanced nations of the world, and thus on concerns about Indian government and managers' abilities to avoid and manage such a crisis, if it were to happen. Vehement demonstrations by the local populace against nuclear energy plants in various locations in India have only increased with time. The issue of risk-benefit trade-off with respect to nuclear power is not a new issue, but the trade-off margins that were socially acceptable prior to Fukushima no longer seem so (Srinivasan and Rethinaraj 2013). The Fukushima nuclear disaster has only aggravated the concerns of those against nuclear power plants. Kudankulam and Jaitapur are two recent examples. Indian judiciary has decided in favour of the Kudankulam plant that has achieved criticality recently, much to the disappointment of the protestors. Chang et al. (2011) analyse two modern-day nuclear power plants, one in Florida and one in Georgia, in the USA and find that there are major risks and cost escalations associated with construction of the two plants which could lead to much higher costs to ratepayers.

Opposition to nuclear plants and significantly negative risk perceptions are not something confined only to India. Some are of the opinion that nuclear power is arguably the riskiest amongst all energy supply technologies (Mills 2012). Germany has decided to phase out nuclear energy completely by 2022. Japan earlier shut down its last nuclear reactor, but has restarted some reactors recently to meet its electricity needs. Interestingly, even in the USA, the revived enthusiasm for nuclear power has dampened after Fukushima. However, all these are high-income countries and do not face the huge energy supply–demand imbalance as faced in India. Each technology, be it renewable or nuclear, is important for bridging the energy gap in India. The challenge in India is to manage the negative risk perceptions around nuclear energy and prioritize investment in the appropriate technology.

10.2 A History of Nuclear Power Generation Costs

Several studies that have analysed historical nuclear cost trends have pointed to the escalating costs for nuclear power plants overtime, and these raise doubts about whether nuclear can become cost competitive (Cooper 2014; Hultman et al. 2007). The 2009 update, of the seminal study (in 2003) on the future of nuclear power, emerging from the stables of MIT, suggests that since 2003 construction costs for all types of large-scale engineered projects have escalated dramatically in the USA.

The estimated cost of constructing a nuclear power plant has increased at a rate of 15% per year (MIT 2009).

In the 2003 MIT study, the authors conclude that 'in deregulated markets, nuclear power is not cost competitive with coal and natural gas. However, plausible reductions by industry in capital cost, operation and maintenance costs and construction time could reduce the gap' (MIT 2003). The authors go on to report that the actual construction costs of nuclear plants completed in the USA during the 1980s and early 1990s compared poorly against estimates at the beginning of these projects. They report that construction schedules experienced long delays, which, together with increases in interest rates, resulted in high financing costs. Emerging regulatory requirements and public opposition to certain projects also resulted in inordinate delays and cost overruns (MIT 2003).

The 2003 report found that capital cost reductions and construction time reductions were plausible, but not yet proven. As of 2009, there was limited evidence that was available to substantiate that such reductions had indeed happened. A study conducted at the University of Chicago in 2004 suggests that first-of-a-kind engineering costs of new nuclear plant designs are likely to increase capital costs in the USA by up to 30% (University of Chicago 2004).

Diesendorf (2011) and Rangel and Lévêque (2015) make the point that the technology levels (and the commensurate safety features) are different across reactor types and over time and could explain higher costs. As a result, a direct comparison of costs is also not easily done. The case of the Kudankulam plant in India, which is the largest plant commissioned in India, is one such where cost comparison against the prevailing costs (at the time) does not make sense. These are the first set of imported LWRs that are being installed in India, and it is early in the learning curve. The domestic know-how and expertise is limited to the large base of PHWRs. Moreover, the agreement between India and Russia for this plant happened in 1988, as part of a larger procurement deal but the construction did not begin until 2000 and the plant was not commissioned until 2013. Over this period, even the reported costs increased by 30% (Subramanian 2013).

A report to the UK Sustainable Development Commission suggests that the 'limited number of published capital cost estimates that apply directly to the UK shows that all appear to derive from studies originally designed to apply to other countries and from vendors of reactor systems'. Given that even in a relatively developed setting where transparency on costs incurred by utilities is more easily accessible, it is certainly an uphill task looking for reliable sources of information on costs on nuclear in the Asian context (Diesendorf 2015).

These issues beget the important question: what is the real cost of nuclear and what are the trends associated with it? International experience suggests that it is not easy to isolate the impact of delays and technology changes, and these will be part and parcel of any renaissance that the nuclear energy sector might witness. Lovering et al. (2016) review a range of studies that explore learning curves and impact on costs and innovation (in the nuclear energy sector) in countries where data has been forthcoming. They argue that a bulk of the past studies has been limited in their scope, focusing primarily on cost trends in the 1970s and 1980s for the USA and France. They suggest that these studies focus only on 26% of the total number of nuclear power reactors completed in the world and only look at two of the 31 countries that generate electricity from nuclear power today.

In order to understand cost trends, they instead carry out a comparison based on reported overnight construction costs (OCCs) from different countries for various reactors that were installed within these countries. The study, the most comprehensive of its kind, spanning six countries, and nearly 58% of global capacity finds that the all countries barring South Korea have experienced a rise in OCCs. Interestingly, the increasing trend has not always been the case and there are phases of development in every country when OCCs have dropped. In two of the Asian countries (Japan and India), nuclear costs have increased but have reached a steady level of sorts in the last phase of construction (400,000 Yen/kW in Japan and USD 2000/kW in India). The latest experience in South Korea, with its standardized design and stable regulatory regime, suggests the possibility of learning by doing in the South Korean industry (Lovering et al. 2016). There is no data that is currently available on costs in China and Russia. It would make for interesting analysis to see whether China has benefited from a Korea like drop in costs on account of indigenization and learning by doing.

Based on the findings in various studies reviewed, we can conclude that rather than rely on learning curves to predict future costs, decision makers should focus on pursuing and developing policies that aim to drive the price of technologies to mitigate GHG emissions. Specifically, policies that drive innovation and better trade and allow for stable regimes in the energy sector will go a long way in incentivizing a domestic industry and drive down long-term costs. Finally, as suggested in the MIT study (2003), reducing the risk premium associated with the technology will go a long way to make nuclear power competitive with other sources (MIT 2003).

10.3 Nuclear Liability Act and Technology Cost

There are significant differences in the valuation of risk across countries, which are reflected in their valuation of liability due to nuclear accidents. The liability amounts are 10 Billion (Bn) US$ for USA, 1500 Million (Mn) Euros for France, 250 Mn US$ for India and unlimited liability in the case of Japan. Moreover, researchers have argued that there is an inherent subsidy over the specified liability amounts. The Price-Anderson Act of 1957 of the USA and Heyes' (2002) subsequent analysis of the implicit subsidy given to nuclear energy operators provide a good basis for one such valuation. Another basis for valuation is the French nuclear liability law.

The Price-Anderson Act (amended) of the USA provides for different levels of protection. The operator carries the primary liability of US$300 Mn (2005 prices) per reactor as the first level. Any damage above this mark is assessed equally against all operators up to about US$96 Mn per reactor. The combined insurance coverage is over 10 Bn USD (American Nuclear Society 2005). This arrangement hence provides the coverage of 10 Bn USD by distributing the liability cost across operators, with no cost to the public exchequer.

The French Nuclear Liability Law, which derives from the Paris and Brussels conventions (1966, amended) like many other nations, allows for three layers of financial liability (Faure and Fiore 2009). As per the revised values, the first tier the operator's liability is capped at 700 Mn Euros, the second layer is a liability of 500 Mn Euros borne by the state and the third layer 300 Mn Euros liability borne by the contracted parties of the French and Brussels convention. These amounts, revised in the Modification Protocols of the Paris and Brussels convention, have not been brought in force yet. The total insurance coverage is hence 1500 Mn Euros per reactor depending on the extent of actual damage. Interestingly, though the overall insurance coverage in the USA is significantly higher than the coverage under the French laws, liability per reactor in the USA is lower than that of a French nuclear reactor.

Moreover, as any nuclear disaster costs billions of dollars which is not necessarily matched by the liability coverage, there is an implicit subsidy involved here. Heyes and Liston-Heyes (1998) estimate this subsidy for USA to be equal to 2.32 Mn USD per reactor per year (1998 prices[1]) of operation. Assuming 45 years of reactor life, total subsidy comes around to 125 Mn USD (in 2005 prices). Faure and Fiore (2009) analyse subsidy for French nuclear reactors for different scenarios of accident costs and probabilities. Their values range from 0.14 Mn Euros to 3.3 Mn Euros per reactor year. Thus for the case with highest expected damages from an accident, the subsidy is almost 150 Mn Euros per reactor, assuming 45 years lifetime.

Comparing the above estimates, it is clear that insurance coverage for a nuclear accident varies from 1500 Mn Euros in France to over 10 Bn US$ in the USA. The insurance financing structure adopted by the USA is interesting as it provides a high level of coverage at the same time not providing a huge financial burden per reactor. The objective of this last statement is not to say that one arrangement is better over the other, but is to highlight the variation in valuation of potential damages due to a nuclear accident, the associated liability coverage and liability financing structures.

In case of India, recent developments indicate that for a high growth trajectory fuelled by nuclear power, India has to continue importing nuclear reactors as indigenous fuel and reactors might not be able to keep up with the pace of nuclear energy demand. The Kudankulam nuclear power plant has light water reactors (LWR) technology imported from Russia. Similarly, India recently concluded high-level negotiations with the USA as well as France, which are potential exporters of nuclear technology for India. The bone of contention was the nuclear liability law, which holds the supplier accountable (in part) for any nuclear accident in India.

---

[1]The price year in this case is to the best of author's understanding as this is not clear from the document.

The 2010 Civil Liability for Nuclear Accident Act (GoI 2010) of India places the responsibility with the operator. Total liability is 2700 Crore INR (450 Mn USD), and operator's liability is limited to 1500 Crore INR (250 Mn USD), beyond which central government will bear the cost. However, the act also gives an opportunity to the operator to have a legal recourse to the supplier if the accident is a consequence of any act of supplier or its employee. As post accident, it is really difficult to prove that the accident occurred due to either equipment fault or due to management negligence, suppliers were resisting as this shifts the financial liability to them even though they are not managing the operations. Russian authorities had earlier stated that a shift of accident liability to Russia will force it to double the price of the light water nuclear reactor exported to India for Kudankulam reactors 3 and 4 (The Indian Express 2012).[2] However, post the nuclear insurance pool arrangement, the contract for 3rd and 4th unit of Kudankulam has been finalized with Russia. Interestingly, the amount of liability is only around 10% of the cost of a nuclear power plant. Still this clause has sparked a high decibel debate. This could be due to a variety of reasons—a concern that this clause will set a precedent for other countries to follow suit, or a future government might simply ignore the liability cap in case of any future accident and impose a higher liability on operators/suppliers (Ramana and Raju 2013). Also, it has been argued that recourse to supplier is a radical departure from international practice and will find difficulty in acceptance worldwide (Jones 2013).

The government unveiled a nuclear insurance pool in June 2015 to address the concerns of suppliers. The Nuclear Power Corporation of India Ltd. (NPCIL) recently got the insurance policy covering all their atomic plants for a risk cover of 1500 crore Rupees for a total premium of Rs. 100 crore per year (The Indian Express 2016). In terms of technology cost hence, it can be robustly concluded that the Indian liability law will not significantly increase the cost of nuclear technology for India. Supplier countries have changed their positions as a consequence of the action taken by Government of India. This was highlighted by Nisha Biswal, a senior US government official, in her recent statement that contours of commercial venture by Westinghouse are taking shape with a deal to provide six reactors to India (Biswal 2016).

10.4 Implications of Increasing Nuclear Power Cost for India's Energy and Climate Policy

The divergence in expectations of future cost varies widely across the stakeholders, the experts typically viewing the risks as small and public perceiving the risks as high. This perceptions gap often narrows during the periods with no major nuclear

[2]Interestingly, as per news reports, the deal for Kudankulam reactors 3 and 4 with Russia is close to finalization as on 11th March 2014. It is unclear from publicly available information if Russia has accepted India's nuclear liability clause as it is.

incidents. However, the gap widens following the major incidents such as at the Fukushima Daiichi Nuclear power plant in March 2011. The divergent views on the external cost of nuclear are also shaped by signals from policy makers. For instance, immediately after the Fukushima incident, the decision to close four nuclear power plants in Germany and phase out the remaining nine nuclear plants by 2022 was viewed as a signal for high risk. A similar interpretation can be made of the Japanese government's decision to shut down all 54 nuclear units in country within a year of the Fukushima incident. Currently, however, the Japanese government has signalled its intent to increase electricity supplied by nuclear power plants. France and Russia were undeterred and continued with their nuclear plans which are indicative of a low nuclear risk perception.

There are alternative ways of managing the opposition to nuclear power plant risks. One possible way is compensating and relocating people residing in the vicinity of proposed sites, something that is challenging to implement given India's prior experience with relocation and rehabilitation in the case of hydro power plants. Another way is by increasing safeguards and risk mitigation measures by deploying nuclear plants with advanced safety features. Another possibility is managing the negative risk by ensuring the local population against any nuclear disaster. This implies that the nuclear operator, or the government as a guarantor, is to provide a risk cover to all the people who are potentially at risk. This does not mitigate the risk, if any, but it does help in assuaging concerns of local residents and dulls opposition.

Irrespective of the risk mitigation choice, one outcome is certain. Any risk mitigation measure is bound to increase the capital and operation costs of any given nuclear power plant. Shukla and Chaturvedi (2012), in an earlier effort, analyse the implications of a targeted approach towards nuclear and solar energy in India. This chapter contributes to the nuclear energy discussion in India from another dimension. Our objective is to present the historical developments on the cost of nuclear power, and highlight implications of increase in cost of operating nuclear power plants for Indian power sector in the short and long term. In other words, we undertake a sensitivity analysis of nuclear energy cost, including scenarios that internalize nuclear accident risk in the technology cost and highlight its implications for India's energy future.

Along with presenting the historical developments on the cost of nuclear power, this chapter takes a century long view of the nuclear energy in India. We follow the scenario approach for the assessment. The scenarios are constructed based on two key factors which would drive the share of nuclear power in the future energy mix.

First is the total cost of nuclear power plant which includes the following: (i) construction costs, (ii) operating costs including fuel cost, (iii) additional measures for addressing externalities and risks during the period of a plant's operating life, (iv) costs of storage of spent fuel and (v) decommissioning costs. The expectation of the future total costs varies widely depending on the respective values assigned to these underlying factors.

Second is the level of global response to climate change. Policy makers around the globe have committed to the 2 °C stabilization target. The Paris Agreement in

2015 aims at achieving temperature increase to 'well below 2 °C'. However, in practice, a gap persists between the emissions pathway to achieve the long-term climate stabilization target and the emissions that the current mitigation policies would deliver (UNEP 2013). The climate responses thus would range between the business-as-usual actions and the policies and measures to 2 °C stabilization.

We consider 12 scenarios for India corresponding to the six (6) varying future trajectories of cost achieved by increasing the overnight construction cost of nuclear power plants relative to the BAU, and two (2) climate scenarios representing 'no climate target' and '2 °C stabilization target'. The paper assesses the change in India's energy mix for these scenarios and examines the sensitivity of nuclear power in the energy mix to the cost of nuclear energy-based electricity generation.

Our cost sensitivity analysis is critical for understanding the potential limitation on increasing the share of this technology as envisaged by the Government of India. The US experience with nuclear power plants has delivered little confidence in the costs and economics of nuclear power plants (Bradshaw and Gruber 2007), and actual overnight capital cost has always overshot analysts' expectations by a wide margin (Cooper 2012) suggesting that growth of nuclear power has been a non-market phenomena. The issues that have affected uncertainty in costs range from design issues to evolving regulatory requirement to long construction periods. Moreover, there are fundamental uncertainties in the cost of key components of fuel cycle (MIT 2011). As highlighted by Kessides (2010), even with equal levelized cost and commercially proven technologies, different risk profiles of different technologies can influence the choice of power generation mix. All these perspectives also provide a motivation for our analysis of the implications of various future cost trajectories for the penetration of nuclear energy in India's power generation mix.

10.4.1 Scenario Description, Cost Assumptions and Modelling Framework

In this context, we present and analyse combination of six nuclear energy technology scenarios and two climate policy scenarios. The technology scenarios range from reference costs, to cost increases of 10, 25, 50 and 100% 2020 onwards relative to the reference scenario (Ref sc), and ultimately to a scenario where nuclear energy is perceived to be prohibitively risky by the society, leading to no new builds of nuclear energy plants implying that the risk-induced costs are too high. These cost assumptions exclude the fuel cost of nuclear energy which depends on the market demand and supply situation and is determined endogenously in the model. Two policy scenarios included here are reference policy scenario with no greenhouse gas mitigation target, and a climate policy scenario that envisages limiting radiative forcing level to 2.6 W/m^2 (proxy for negotiated 2 °C temperature stabilization target) in 2100. The below table details the scenario description and cost assumptions (Table 10.1).

We use the modelling framework of Global Change Assessment Model (GCAM, Indian Institute of Ahmedabad (IIMA) version), for understanding the short-, medium- and long-term implications of incorporating risk valuations into nuclear energy generation costs for India. GCAM is an energy sector-focused integrated assessment model with an energy module, land use module and a climate module within the same framework. GCAM has been widely used for global and regional energy and climate policy scenario exercises (refer Clarke et al. 2008; Calvin et al. 2009; Wise et al. 2009; Kyle and Kim 2011; Eom et al. 2012; Shukla and Chaturvedi 2012; Chaturvedi et al. 2013; Edmonds et al. 2012). The world within GCAM-IIMA is divided into 14 regions, with India as a separate region. The

Table 10.1 Scenario description and cost assumptions

Scenario name	Scenario description	Nuclear power plant levelized capital and O&M costs (in 2012 INR per kWh)		
		2020	2050	2095
Ref	Scenario with no climate policy targets and reference nuclear energy cost	5.25	4.49	4.01
Ref_10%	Scenario with no climate policy targets and non-energy nuclear cost 10% higher compared to Ref sc	5.78	4.94	4.41
Ref_25%	Scenario with no climate policy targets and non-energy nuclear cost 25% higher compared to Ref sc	6.56	5.61	5.01
Ref_50%	Scenario with no climate policy targets and non-energy nuclear cost 50% higher compared to Ref sc	7.88	6.74	6.02
Ref_100%	Scenario with no climate policy targets and non-energy nuclear cost 100% higher compared to Ref sc	10.50	8.98	8.02
Ref_Retire	Scenario with no climate policy targets and no new builds of nuclear energy plants	No new nuclear power plants built		
2.6	Scenario with climate policy target of 2.6 W/m^2 radiative forcing by 2100 and reference nuclear energy cost	5.25	4.49	4.01
2.6_10%	Scenario with climate policy target of 2.6 W/m^2 radiative forcing by 2100 and non-energy nuclear cost 10% higher compared to Ref sc	5.78	4.94	4.41
2.6_25%	Scenario with climate policy target of 2.6 W/m^2 radiative forcing by 2100 and non-energy nuclear cost 25% higher compared to Ref sc	6.56	5.61	5.01
2.6_50%	Scenario with climate policy target of 2.6 W/m^2 radiative forcing by 2100 and non-energy nuclear cost 50% higher compared to Ref sc	7.88	6.74	6.02
2.6_100%	Scenario with climate policy target of 2.6 W/m^2 radiative forcing by 2100 and non-energy nuclear cost 100% higher compared to Ref sc	10.50	8.98	8.02
2.6_Retire	Scenario with climate policy target of 2.6 W/m^2 radiative forcing by 2100 and no new builds of nuclear energy plants	No new nuclear power plants built		

Source Authors' analysis

Table 10.2 GCAM-IIM economic and demographic assumption

	GDP	Population
	2005 Bn US$ (MER)	Bn
2005	748	1.130
2020	2306	1.347
2035	6860	1.568
2050	16,179	1.736
2095	33,671	1.512

Source Authors' analysis

strength of GCAM is the detailed representation of the energy sector, both on the supply and demand sides. GCAM models energy demand for three end-use sectors —building sector, industry sector and transportation sector. On the supply side, electricity production is modelled in detail with nine fuels competing for electricity production, with more than one technology within each fuel. Nuclear energy, being a zero-carbon source, is a key technology for electricity generation in GCAM.

The IIM Ahmedabad version of GCAM is different from core GCAM in its demographic and economic growth assumptions and representation of the building sector. GCAM-IIM has higher per capita income growth assumptions, and the building sector in GCAM-IIM is further disaggregated into urban residential, rural residential and commercial building sector. Also, we revise the cost of all key power generation technologies for India and the world, based on developments in the Indian power sector as well as International Energy Agency cost outlook. The demographic assumptions are given in Table 10.2. For earlier studies using GCAM-IIM, please refer Shukla and Chaturvedi (2012, 2013), Chaturvedi and Shukla (2013), Chaturvedi et al. (2014).

10.4.2 The Role of Nuclear Energy in the Reference World for India

India's electricity generation is bound to witness a high rate of growth given the energy-starved situation today. This growth will be particularly high post 2020, when Indian average per capita incomes will drive rapid move towards electricity-based technologies for meeting building energy service requirements, mainly cooling and appliances. Some move will also be towards electricity-based private and public transportation modes and more efficient energy use processes in Indian industry. In our Ref sc (or business-as-usual scenario), electricity generation increases to 2700[3] Bn kWh by 2030 and further doubles by 2050. This is compared to 1175 Bn kWh electricity generated in 2013–2014.

[3]Our estimate for future electricity generation depends on economic growth rate, rate of electrification of end-use sectors, energy efficiency improvements, and transmission & distribution losses. Our estimates could increase by 15–20% or even decline by 8–10% if the assumptions are changed.

In terms of the character of electricity generation, there is not a significant shift observed between now and 2050, and even beyond. A large part of India's electricity generation is based on coal. This does not mean that there is little increase in non-fossil energy. Both solar and wind continue to grow. But the share of hydro power diminishes largely because of non-economic factors such as sociopolitical constraints. Overall, coal still dominates India's power generation mix under the BAU scenario even in the long run (Fig. 10.1a).

(a)

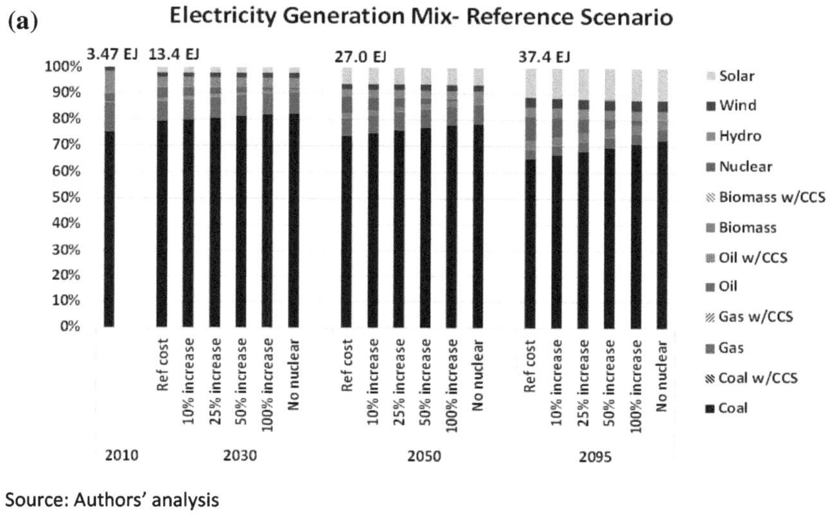

(b)

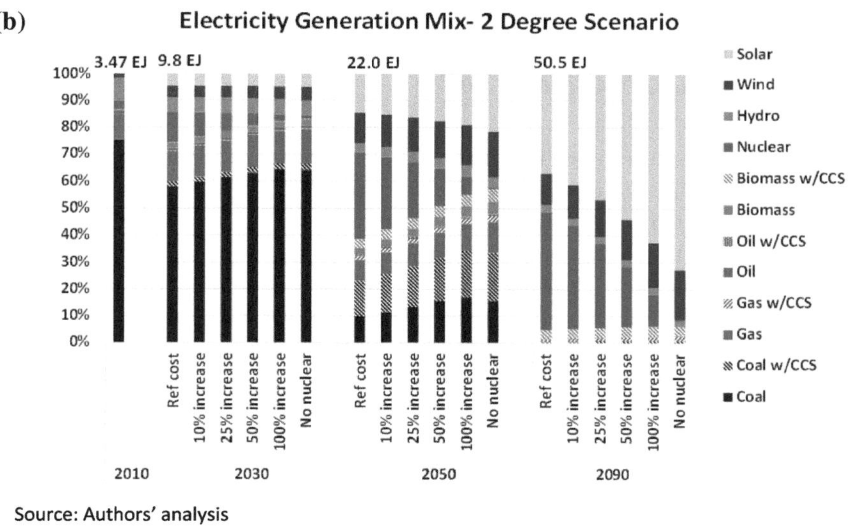

Fig. 10.1 **a** India's electricity generation mix under reference scenario. **b** India's electricity generation mix scenario under 2 degree scenario

India has ambitious plans to increase the penetration of nuclear energy in its power generation mix. The target is to increase the penetration to 63 GW in 2032. The cost of nuclear power plants will however determine the penetration of nuclear energy in India to a great extent. Based on understanding of the cost of electricity production from imported light water reactors in recent discussions, we assume that cost of nuclear power-based electricity generation in 2020 will be 5.25 Rs/kWh in terms of 2012 prices. We assume that with better learning and understanding of this technology, there will be a 15% decline in this cost in real terms between 2020 and 2050, and further 10% decline between 2050 and 2100.

With the assumed relative costs, we see that nuclear power grows fourfold between 2010 and 2030, even though its share in India's electricity generation remains largely stable due to an overall increase in India's electricity generation. This implies that even after an increased cost of power production as what is being observed and predicted, nuclear energy penetration will only increase in India in terms of production capacity. The rate of growth is, however, much less as compared to the growth observed in solar-based electricity generation due to rapidly declining cost of solar energy. Still, in the long run, both nuclear and solar energy are established as two key energy technologies for India after fossil energy sources.

Reference scenario by definition is devoid of any climate policy-specific intervention and should be only regarded as a point of departure. This scenario only shows where Indian power generation mix would be heading in absence of any climate policy intervention. In reality however, we are observing the Indian policy makers pushing both solar and wind, as well as nuclear energy in India's power generation mix with an aim of moving away from a heavily coal-based future.

10.4.3 The Role of Nuclear Energy in India's Climate Policy

Climate policy scenario with reference nuclear costs (2.6 Scenario) witnesses a decrease in electricity generation in the short run, 27% by 2030 relative to Ref sc due to a sudden increase in carbon intensive energy prices. For attaining the 2-degree target, the endogenously estimated global carbon price increases from 28 US\$/tCo_2 (2010 prices) in 2020 to 122 US\$/tCo_2 in 2050 and 1397 US\$/tCo_2 in 2100. The increase in the carbon prices shows the marginal value of each unit of carbon dioxide that needs to be removed to attain the 2-degree target. It also shows that as potential mitigation opportunities get exhausted, mitigating an additional unit of carbon dioxide will be much more expensive than mitigating the last unit.

Beyond 2030 however, more and more end-use services move towards electricity, and by 2050, electricity generation is almost 35% higher relative to the Ref sc. Increased electrification is a must for cost-effective climate policy as there are multiple options to reduce carbon emissions from the electricity generation sector. Reducing direct emissions from the end-use sectors is much more costly. Shifting end-use emissions to the electricity generation sector as a first step, and then mitigating emissions from this sector is an important strategy for meeting global climate mitigation goals.

India's electricity generation system will need to move out of the fossil energy sources in a big way to attain the deep mitigation targets. In this context, the criticality of nuclear energy-based electricity for India becomes obvious under the stringent climate policy scenario (Fig. 10.1b). This energy source powers on to take 11% share in 2030, 32% share in 2050 and 44% share[4] in India's generation portfolio by the century end. This implies that nuclear energy has the biggest impact compared to other low-carbon sources in India's electricity generation strategy. Solar energy takes a share of 14% in 2050, beyond which it keeps growing to take an ever larger share. Even carbon capture and storage (CCS) including biomass with CCS takes 18% share in 2050, though its share reduces dramatically post 2050 with increasing share of nuclear and solar energy. Thus by the end of century, coal is almost completely displaced from India's electricity production portfolio and nuclear energy is clearly established as a key source for powering India's electricity generation capacity under a climate constrained world.

10.4.4 What if Nuclear Generation Cost Increases Further?

This section explores the implications of cost increase across the range of 10–100% (relative to Ref sc) in the cost of nuclear-based electricity. Even a 10% further increase in the cost of nuclear energy leads to a 20% decline in its penetration in 2030 and beyond compared to the Ref sc. Though installed capacity of nuclear energy declines by 20% in the medium to long term, the overall character of India's electricity mix is similar to nuclear energy still providing 5% of electricity generation in 2050 and 7% in 2090. We see a similar effect of a 10% cost increase under the climate policy scenario, and nuclear energy still takes 28% share in 2050 and 39% share in 2090. We can hence say that a 10% increase in electricity generation cost by nuclear energy relative to the Ref sc costs will not have any significant impact on India's electricity generation scenarios. This technology will still remain the focal point for India's response to emission mitigation challenges. It should be highlighted here that the Ref sc assumes a decline in costs of all technologies in the future.

A higher increase of 25–50% of nuclear technology's cost will have significant implications for the next 15–20 years irrespective of the policy scenario. Due to

[4]Share of respective energy sources in our model depend on their relative costs and technical characteristics. If the relative costs change, then there will be a different outcome. e.g. if the cost of solar PV declines further, we will see a lower share of nuclear and an even higher share of solar under the climate policy scenarios. Rapidly changing costs imply that the quantitative scenario outcomes will change with every update in the analysis, however the broad economic insights from this analysis will still hold.

increase cost, installed nuclear energy generation capacity declines by 36–55% under no climate policy scenario and 47–64% under climate policy scenario by 2030 in the near term. In the longer run however, it is the climate policy regime which critically determines the penetration of this technology. Under the Ref sc without any mitigation targets, nuclear energy share declines to 2–4% in 2050. However, a carbon price ensures that nuclear energy plays a critical role, with 22–31% electricity still produced by this technology in 2090, up from 13–19% in 2050.

A 100% increase in nuclear energy cost (or 10.5 Rs./KWh in terms of 2012 prices) post 2025 is an extreme scenario and reflects further increases in the cost of nuclear energy with higher safety features, increases in the underlying cost of materials used for building nuclear power plant and even increase in costs like that for decommissioning. This extent of increase in risk-induced cost of nuclear energy will have serious implications for the share of this technology under a reference scenario world, and its share declines to 1% in 2050 and 2% in 2090. Post 2030, installed capacity relative to the Ref sc declines by 72–82% for all future years.

Similar effect is observed under the climate policy scenario as well. First, nuclear energy does decrease significantly due to doubling of cost. In 2050, the decline in its share is from 32% in 2-degree scenario with reference cost to 6% when costs double. For 2090, the decline is from 44 to 12%. Thus, in the long run, a doubling of non-energy cost of nuclear energy does decrease its importance as a critical technology in India's emission mitigation strategy. Still this source plays an important role in meeting the twin goals of meeting energy demands as well as climate policy targets. Second, two key technologies gain in share as a result of decrease in the share of nuclear energy. Solar energy increases its share to almost two-thirds of total power generation in 2090 with a doubling of nuclear energy cost. Wind energy share also increases, though it is still far less than that of solar energy.

10.4.5 Can India Think About Becoming a No Nuclear Energy Country?

Under a reference scenario, with no obligations towards emission mitigation and adaptation, India can aim towards a no nuclear future. Achieving this objective is definitely possible from a technology perspective. Without any climate constraints, India can move towards either fossil energy or renewable sources of energy. However, cost dynamics ensure that even in the long run, nuclear is replaced by fossil sources, mainly coal rather than renewable energy. Unless a renewable energy push policy is pursued aggressively, fossil energy will dominate Indian energy systems.

This move towards even greater reliance of India's electricity production on fossil energy has implications for carbon dioxide emissions from the country. A no nuclear scenario will result in a substantial increase in emissions from the power generation sector. Under the Ref sc, emissions from India's power generation increase by 12–15% post 2050 when nuclear energy is retired. This directly implies that India will further move away from the goal of emission mitigation. In other words, reaching emission mitigation targets in the future will be more challenging and costly to achieve.

The important point to be noted here is that the stabilization scenario is a global stabilization regime, and as emission mitigation in India and elsewhere in the world is now more costly in the electricity generation sector (one key zero-carbon technology is absent from the mitigation portfolio), mitigation is shifted to some other sectors and regions where it is cheaper. This leads to a change in distribution of the global emission mitigation burden, but the overall character of this distribution along with global emissions remains unchanged.

A key outcome of increased cost of or phasing out of nuclear technology for India is steep increase in the cost of meeting climate policy emission mitigation objectives. There is almost no change in the short- and medium-term mitigation costs, however long-term costs change significantly. From 2060 onwards, we see a steep increase in additional mitigation cost due to removal of nuclear technology. It is thus clear that even a 25 or 50% increase in the cost of nuclear energy relative to reference cost means a lot in terms of long-run mitigation cost for achieving a stringent climate target. A complete phase out will make it even costlier unless there is significant decline in the long-run cost of alternative low-carbon technologies.

10.5 Climate Liability Versus Nuclear Liability

Two important factors have traditionally driven the policy in favour of nuclear power generation in India. The first is the goal of universal access to electricity and clean energy at affordable prices. This goal is directed at alleviating the energy poverty that grips millions who either lack the connectivity to the electricity grid or who face frequent and long-duration power blackouts. This implies that India has to ramp up power production capacity to meet the latent demand of the underserved masses as well as the growing future demand of those with rising incomes. Our analysis shows that optimal electricity generation mix for India, across diverse scenarios, includes varying mix of renewable energy, fossil energy and nuclear. The hurdles to renewable options include high cost of initial deployment, intermittency associated with the generation and lack of clarity on the ultimate potential. As far as fossil fuels are concerned, the pace of resource extraction has suffered from numerous legal and policy hurdles and has left many power-generating assets stranded. Nuclear is then seen by policy makers, even under the business-as-usual scenario, as a viable third pillar of energy supply that can be ramped up to bridge the electricity demand–supply gap.

The second important factor that concerns policy makers is the long-run energy security. As our modelling results show, the energy demand in India is likely to exhibit persistently high growth rates. Given the relatively meagre endowments of oil and gas, India's import of these energy commodities and the impact on current account deficit are likely to follow a similar trajectory as the demand. India is endowed with significant coal resources, but the production has lagged due to procedural and technological lock-ins, besides lack of project clearances due to genuine environmental concerns. As a result, India has witnessed increasing imports of coal in the last decade though there has been a recent decline in coal imports. India is the one of the few countries to have pursued the nuclear program (even prior to developing nuclear weapons) in a steadfast manner. From the early days of Independence, the country expended significant resources in developing the necessary institutions and human resource to run a large-scale civilian nuclear power program. The fact that the Department of Atomic Energy is directly under the charge of the Prime Minister of the country underscores the importance that was accorded to it by the founding fathers who recognized the energy security implications of pursuing a nuclear program. Nuclear power presents an alternative route to achieving energy security. The marginal cost of nuclear power is significantly lower than that from conventional thermal power sources and comes without the large macroeconomic risks associated with the price of imported fossil fuels. A nation with a large domestic electricity market and having capacity to develop own nuclear power plant industry can find nuclear as a ready answer to energy security concerns. India's Atomic Energy Commission's ambitious long-term nuclear strategy is to construct 500 GW of largely indigenously developed nuclear capacity in next fifty years (World Nuclear Association 2014). Nuclear, in this context, can de-couple domestic energy prices to a large extent from the risks associated with the uncertainties of global energy markets.

Besides these two factors, our research highlights a third emerging factor that is now providing fillip to the arguments in favour of nuclear energy in India. This third factor is climate change mitigation. The results of our modelling analysis clearly show that India's greenhouse gas emissions would rise significantly, under all reference scenarios, by the end of the century. India has a huge population which shall continue to depend primarily on, agricultural, forestry and fisheries for livelihood for next several decades, and India's diverse and unique ecosystems are highly vulnerable to climatic changes (Shukla et al. 2003). India has actively participated in the shaping of the Paris Agreement and has endorsed 2 °C global climate stabilization target. India's National Action Plan on Climate Change (GoI 2008) has proposed various missions to initiate and implement early actions to align this climate target with national economic development plans, policies and programmes.

Nuclear energy fits well with the vision to offset the greenhouse gas liability. However, fallouts from nuclear accidents (large or small) bring in a different liability. Our analysis shows that under current expectations of nuclear-based electricity cost, India's optimal electricity mix for meeting emission mitigation concerns shall have nearly 45% share of nuclear energy at the end of the century. This means building nuclear power plants at many locations and many units at each location.

Historically, the largest nuclear power plant installation has been Kashiwazaki-Kariwa nuclear power plant in Japan with 8.2 GW capacity. India's proposed nuclear plant at Jaitapur has a 10 GW capacity. In case of 45%

penetration scenario mentioned above, India will need 25 nuclear power plant sites in 2050 and 75 nuclear power plant sites of the size of Jaitapur spread across India in 2090. Thus, if the current economic expectations were to continue to drive the penetration of nuclear, and assuming that there are no institutional and other issues limiting deployment of nuclear in India, nuclear energy will become the focal point of India's response to climate change mitigation challenges. In this context, increased nuclear liability effectively offsets climate change liability. Nuclear risks one could argue are more clearly understood and can be planned for, as opposed to the nascent understanding of global climate change and the far reaching impacts on livelihoods, human health and the broader ecosystem.

However, in case costs double, nuclear electricity share will decrease dramatically in the near as well as long term. In this case, renewable energy and CCS will be technologies delivering most of the greenhouse gas mitigation. This brings in the role of these 'horizon technologies' and their interactions with nuclear technology for meeting India's energy and emission challenges. Our scenarios include carbon capture and storage (CCS) as an important horizon technology that contributes to emission mitigation. However, given that CCS is a commercially unproven technology, uncertainty around this option is high. Our results show that if CCS becomes viable and acceptable as per current expectations, it will be employed in tandem with coal in the next two decades and with commercial biomass in the longer run. However, in case CCS is not proven commercially successful, the greater burden of emissions mitigation will fall on renewables and nuclear. In such a scenario, the nuclear will have even more vital role in meeting emission mitigation.

10.6 Conclusion: Moving Ahead with a Wider Understanding of Nuclear Energy Cost

India's energy policy makers have viewed nuclear energy as a vital option for meeting the growing electricity demand. The general consensus in the political establishment on the issue is evident from the signing of civil nuclear cooperation agreements with a host of countries including USA, Russia and France on fuel and technology supply, and the development of a nuclear liability framework. In civil society though there are divergent views on India's nuclear energy ambitions, given the potential risks associated with nuclear power. In case of nuclear energy, wide perception gap exists in India among the stakeholders about its external cost. This perception translates into a wide range of external costs of nuclear energy from risk mitigation measures and the residual liability costs.

A review of the historical experience with the overnight capital cost of nuclear power plants suggests that costs have been increasing with time. However, it is not right to compare a 30- or even 15-year technology with today's technology because of many different features such as enhanced passive safety measures. Interestingly, in all regions, there have been periods where in the OCC of reactors has declined as well. Within India, there is not yet a rich experience of imported light water reactors and hence cost projections are tricky at best. Understanding the implications of potential further increases in the cost of nuclear energy is critical for Indian planners and policy makers. This paper is an attempt to understand the historical increase in nuclear energy cost, and explore the sensitivity of India's long-term energy and climate change mitigation policy to further increases in nuclear energy cost.

To this end, we analyse a suite of climate policy and nuclear energy technology cost scenarios within the integrated assessment modelling framework of GCAM. Our scenarios are constructed on two parameters that may deeply affect the future energy system in India: I) the price of nuclear technology and II) the future climate change regime. We consider six nuclear technology costs regimes (ranging from reference cost to a complete nuclear shutdown scenario) and two climate regimes (no climate policy and 2 °C stabilization policy).

We find that with current expectations of nuclear-based electricity cost, which is almost double of electricity cost from coal-based power plants, nuclear energy is competitive even in the absence of global climate stabilization policies. However, if costs further increase, then the increase in nuclear energy deployment will be seriously curtailed in the short as well as long run. Economics then favours increased deployment of fossil fuels, leading to significant increase in carbon dioxide emissions. Lower nuclear energy penetration will hence increase the quantum of emissions that need to be mitigated for meeting potential emission targets in the future.

The trend highlighted above also holds under a climate policy world, wherein the nuclear energy penetration declines significantly under high-cost scenarios. However, our results show that nuclear energy is competitive, in the long run, even with high costs if global climate stabilization policies corresponding to global 2 °C stabilization target are pursued. Nuclear energy can make important contribution to India's energy security, through the century, under all climate stabilization scenarios. This clearly highlights the criticality of nuclear technology in India's long-term emission mitigation strategy. Given the fact that Government of India is actively engaged in global climate negotiations and the 2 °C stabilization target is an accepted target by the Government of India (GoI 2008), our suggestion is to keep concurrent focus on nuclear energy as well as climate change mitigation while presenting the risks. Interestingly, the peak year for Indian emissions under the climate policy scenarios shifts to 2050, a shift of 15 years towards the future, when the nuclear technology costs double or it is retired from the energy mix portfolio.

Increased costs of nuclear energy have serious implications for costs of meeting emission mitigation targets in the long run. In the medium run, till 2050, other low-carbon technologies come in place of nuclear and the effect on mitigation cost is minimal. However in the second half the century, as the initial low-cost potential of solar energy and CCS gets exhausted, mitigating emissions becomes increasingly expensive in the wake of higher cost or absence of nuclear technology. Abatement cost for India in 2090 as a percentage of GDP is 3.1% with reference to nuclear energy cost and penetration, which increases to 5.1% when nuclear energy technology is completely phased out.

The key energy policy insight from our analysis is that Indian policy makers need to keep a continuous and close watch on the cost of nuclear power plants, and try to find innovative ways of reducing cost to achieve the targeted share of nuclear energy in India's energy mix. One potential way is to domestically manufacture the labour-intensive components of the nuclear power plants. Greater reliance on a domestic technology will also ensure at least to some extent that the cost of nuclear power is under control compared to that of imported technology.

Though the results of our study show that taking on nuclear liability is important for offsetting the climate liability, we should also highlight the need of transparent and independent regulation of nuclear power plants. The regulator should not only be independent as per legal requirements, but should also be perceived as independent and powerful to act against operators if required for addressing concerns of the local population. Such a transparent regulator must place public interest and not utility interest at the forefront. Indian nuclear energy policy should have a special focus on detailed response plans and readiness for each nuclear power plant site, and management plans for disaster preparedness and evacuation in case of any nuclear accident. There should be resources dedicated upfront for better disaster preparedness and response, which will help in addressing the concerns of people.

Overall, the scenario assessment shows that the fraction of nuclear energy in India's future optimal energy mix is sensitive to cost increases and strictness of climate stabilization target and related global policies. The debates and arguments remain on what position India should take for future deployment of nuclear energy power plants. The stakeholders having higher risk perception who argue for stopping all nuclear power plants can in fact show that sacrificing 2% of GDP in 2090 is a little cost for doing away with the potentially disastrous nuclear incidents. On the other extreme, the stakeholders having lower risk perception of nuclear energy will argue that with improved safety features safeguards and inherently safer new generation of nuclear power plants, exclusion of the nuclear option will add significantly to the energy cost in India in all scenarios and more so in the stringent climate stabilization scenario. To them, lower cost of electricity would enhance welfare of millions who currently lack electricity access and bear the cost of entrenched energy poverty. There is merit to both arguments, and the optimal response would have to address the perception gap about the risks of nuclear power among stakeholders, and concomitantly provide a credible and transparent analysis of the social benefits and risks. Finally, bridging the information chasm about the risks from nuclear power plants among stakeholders will help to find best fit for nuclear power in India's long-term optimal energy mix.

Acknowledgements We thank Dr. Anil Kakodkar (Former Chairman, Atomic Energy Commission of India and Former Director, Bhabha Atomic Research Centre, India) for his comments on the cost sensitivity scenario analysis undertaken in our paper and making our discussions more relevant to Indian energy and nuclear policy. We also thank Dr. Jae Edmonds and Dr. Leon Clarke (Joint Global Change Research Institute/Pacific Northwest National Laboratory, USA) for providing access to Global Change Assessment Model (GCAM).

References

Abraham, I. (2011). The risks of nuclear power. *Economic and Political Weekly* (16) (EPW commentory).

American Nuclear Society. (2005). *Price-Anderson act. Background for position statement 54.*

Bhaskar, U. (2015). *India's per capita electricity consumption touches 1010 kWh.* Livemint. http://www.livemint.com/Industry/jqvJpYRpSNyldcuUlZrqQM/Indias-per-capita-electricity-consumption-touches-1010-kWh.html. Accessed on 2 Dec 2015.

Biswal, N. (2016). *Security and strategic outcomes from the modi visit. remarks by Nisha Biswal.* Assistant Secretary, Bureau of South and Central Asian Affairs, US Department of State. http://www.state.gov/p/sca/rls/rmks/2016/258401.htm. Accessed on 2 Dec 2016.

Bradshaw, G. B., Gruber, C. O. (2007). *New nuclear power plants—Learning from history to understand costs and mitigate risks.* 2007 AACE International Transactions.

Calvin, K., Edmonds, J., Bond-Lamberty, B., Clarke, L., Kim, S. H., Kyle, P., et al. (2009). 2.6: Limiting climate change to 450 ppm CO_2 equivalent in the 21st century. *Energy Econ, 31,* S107–S120.

Chang, M., White, D., Hausman, E., Hughes, N., Biewald, B. (2011). *Big risks, better alternatives: An examination of two nuclear energy projects in the US.* Synapse Energy Economics, Inc.

Chaturvedi, V., Eom, J., Clarke, L., & Shukla, P. R. (2014). Long term building energy demand for India: Disaggregating end use energy services in an integrated assessment modeling framework. *Energy Policy, 64,* 226–242.

Chaturvedi, V., Kim, S., Smith, S. J., Clarke, L., Yuyu, Z., Kyle, P., et al. (2013). Model evaluation and hindcasting: An experiment with an integrated assessment model. *Energy, 61,* 479–490.

Clarke, L., Krey, V., Weyant, J., & Chaturvedi, V. (2012). Regional energy system variation in global models: Results from the Asian modeling exercise scenarios. *Energy Economics, 34,* S293–S305.

Clarke, L., Wise, M., Edmonds, J., Placet, M., Kyle, P., Calvin, K., Kim, S., Smith, S. (2008). CO_2 emissions mitigation and technological advance: An updated analysis of advanced technology scenarios. Pacific Northwest National Laboratory Technical Report PNNL-18075. Richland, WA, USA: U.S. Department of Energy.

Cooper, M. (2012). Nuclear safety and affordable reactors: Can we have both? *Bulletin of the Atomic Scientists, 68*(4), 61–72.

Cooper, M. (2014). *The economics of nuclear power.* In 'Corporate Knights', Winter 2014 issue, 3 Feb 2014. http://www.corporateknights.com/channels/utilities-energy/the-economics-of-nuclear-power-13914427/. Accessed on 5 Nov 2015.

Diesendorf, M. (2011). *Nuclear power and energy security in East Asia.* Edited Series: Routledge.

Diesendorf, M. (2015). The economics of nuclear energy. In R. Basrur & K. S. L Collin (Eds.), *Nuclear power and energy security in Asia* (pp. 50–71).

Edmonds, J. A., Calvin, K., Clarke, L., Kyle, P., & Wise, M. (2012). Energy and technology lessons since Rio. *Energy Economics, 34,* S7–S14.

Eom, J., Clarke, L., Kim, S., Kyle, P., & Patel, P. (2012). China's building energy demand: Long term implications from a detailed assessment. *Energy, 46,* 405–419.

Faure, M., & Fiore, K. (2009). An economic analysis of nuclear law subsidy. *PACE Environmental Law review, 26* (2).

GoI (Government of India). (2008). *National action plan on climate change.* India: Prime Minster'sCouncil on Climate Change, Government of India.

GoI (Government of India). (2010). *The Civil Liability for Nuclear Damage Act 2010.* Act of Indian Parliament.

GoI (Government of India). (2011). *Low carbon strategies for inclusive growth: An interim report.* India: Planning Commission, Government of India.

Heyes, A. (2002). *Determining the price of Price-Anderson.* Regulation, Winter 2002–03.

Heyes, A.G., & Liston-Heyes, K. (1998). Subsidy to nuclear power through Price-Anderson liability limit: Comment. *Contemporary Economic Policy, 16*, 122–124.
Hibbs, M. (2012). Nuclear energy 2011: A watershed year. *Bulletin of the Atomic Scientists, 68*(1), 10–19.
Hultman, N. E., Koomey, J. G., & Kammen, D. M. (2007). What history can teach us about the future costs of U.S. nuclear power. *Environmental Science & Technology, 41*(7):2087–2094.
IAEA (International Atomic Energy Agency). (2012). *Nuclear technology review 2012*. Vienna: International Atomic Energy Agency.
IEA (International Energy Agency). (2015). World energy outlook 2015. Paris, France: OECD/IEA.
Jones, T. (2013). Why India's nuclear liability law is harming Indian interests. *NEI Nuclear Notes*, 18 Oct 2013. http://neinuclearnotes.blogspot.in/2013/10/why-indias-nuclear-liability-law-is.html. Accessed on 10 Jan 2014.
Kessides, I. N. (2010). Nuclear power: Understanding the economic risks and uncertainties. *Energy Policy, 38*, 3849–3864.
Kour, V., & Dar, S. S. (2013). Significance of energy security in 21st century with reference to nuclear energy scenario of India. *IOSR Journal of Humanities and Social Science, 10*(5), 43–50.
Kyle, P., & Kim, S. H. (2011). Long term implications of alternative light duty vehicle technologies for global greenhouse gas emissions and primary energy demands. *Energy Policy, 39*, 3012–3024.
Lovering, J. R., Arthur, Yip, & Ted, Nordhaus. (2016). Historical construction costs of global nuclear power reactors. *Energy Policy, 91*, 371–382.
Mallah, S., & Bansal, N. K. (2010). Allocation of energy resources for power generation in India: Business as usual and energy efficiency. *Energy Policy, 38*(2), 1059–1066.
Mills, E. (2012). Weighing the risks of climate change mitigation strategies. *Bulletin of the Atomic Scientists, 68*(6), 67–78.
MIT (Massachusetts Institute of Technology). (2003). *The future of nuclear power—An interdisciplinary study*.
MIT (Massachusetts Institute of Technology). (2009). *Update to the The future of nuclear power study (2003)—An interdisciplinary study*.
MIT (Massachusetts Institute of Technology). (2011). *The future of the nuclear fuel cycle. An interdisciplinary MIT study*.
Mohapatra, D. K., & Mohanakrishnan, P. (2010). A methodology for the assessment of nuclear power development scenario. *Energy Policy, 38*(8), 4330–4338.
Ramana, M. V. (2009). The Indian nuclear industry: Status and prospects. Nuclear energy futures papers. The Centre for International Governance Innovation.
Ramana, M. V. (2012). *The power of promise: Examining nuclear energy in India*. New York: Viking Publication.
Ramana, M. V., & Raju, S. (2013). The impasse over liability clause in Indo-U.S. nuclear deal. *The New York Times*, 15 Oct 2013. http://india.blogs.nytimes.com/2013/10/15/the-impasse-over-liability-clause-in-indo-u-s-nuclear-deal/?_php=true&_type=blogs&_r=2. Accessed on 10 Jan 2014.
Rangel, L. E., & Leveque, F. (2015). Revisiting the cost escalation curse of nuclear power: New lessons from the French experience. *Economics of Energy & Environmental Policy, 4*(2).
Remme, U., Trudeau, N., Graczyk, D., & Taylor, P. (2011). *Technology development prospects for the Indian power sector. IEA information paper*. Paris, France: OECD/IEA.
Shukla, P. R., & Chaturvedi, V. (2012). Low carbon and clean energy scenarios for India: Analysis of targets approach. *Energy Economics, 34*, S487–S495.
Shukla, P. R., & Chaturvedi, V. (2013). Sustainable energy transformations in India under climate policy. *Sustainable Development, 21*, 48–59.
Shukla, P. R., Sharma, S. K., Ravindranath, N. H., Bhattacharyya, S., & Garg, A. (Eds.). (2003). *Climate change and India: Vulnerability assessment and adaptation*. Hyderabad: Universities Press.

Sokolski, H. (2010). The high and hidden costs of nuclear power. *Policy Review*.
Srinivasan, T. N., & Rethinaraj, Gopi T. S. (2013). Fukushima and thereafter: Reassessment of risks of nuclear power. *Energy Policy, 52,* 726–736.
Subramanian, T. S. (2013). Ready to run. *Frontline Magazine*. http://www.frontline.in/the-nation/ready-to-run/article4944731.ece. Accessed on Aug 15 2016.
The Indian Express. (2012). *Reactors to cost double: Russia*. In print on 20 Dec 2012, New Delhi Edition. http://archive.indianexpress.com/news/reactors-to-cost-double-russia/1047796/on. Accessed on 25 Sept 2013.
The Indian Express. (2016). *Policy issued three weeks ago, nuclear plants in India get insurance cover for first time*. In print on 21 Aug 2016, New Delhi Edition. http://indianexpress.com/article/business/business-others/policy-issued-3-weeks-ago-nuclear-plants-in-india-get-insurance-cover-for-first-time-2859715/.
UNEP (United Nations Environment Program). (2013). *The emissions gap report 2013*. A UNEP Synthesis Report.
University of Chicago. (2004). *The economic future of nuclear power*.
Wise, M. A., Calvin, K. V., Thomson, A. M., Clarke, L. E., Bond-Lamberty, B., Sands, R. D., et al. (2009). The implications of limiting CO_2 concentrations for land use and energy. *Science, 324,* 1183–1186.
World Nuclear Association. (2014). *Nuclear power in India*. http://world-nuclear.org/info/Country-Profiles/Countries-G-N/India/#.UnIjeOK3Vio.

Author Biographies

Dr. Vaibhav Chaturvedi is a Research Fellow with the Council on Energy, Environment and Water (CEEW). His expertise is on the topic of energy and climate change with a special focus on modeling long term energy and emission pathways, and he has been instrumental in strengthening CEEW's collaboration with premier global institutions. Dr. Chaturvedi has been reviewing for several leading international journals in the domain of energy and climate policy. Prior to joining CEEW he worked with Pacific Northwest National Laboratory (USA). Dr. Chaturvedi has received his doctorate from Indian Institute of Management (Ahmedabad) in Economics and Masters from the Indian Institute of Forest Management. He has published extensively in reputed publications worldwide (Dr. Chaturvedi can be contacted at: vaibhav.chaturvedi@ceew.in).

Prof. Priyadarshi R. Shukla has worked as an Assistant Professor (1979–1988), Associate Professor (1988–1991) and Professor (1991–2015) at the Indian Institute of Management, Ahmedabad, India. Prof. Shukla is currently the co-chair of Working Group III of the IPCC. He is in charge of coordinating analysis of the mitigation of climate change for the IPCC's Sixth Assessment Report. He has been associated with IPCC as an author since 1993—as a lead author of IPCC WGIII SAR (Chapters 8, 9 and 10), TAR (Chapter 8), AR4 (Chapter 2) and AR5 (Chapter 6). He is also a lead author of IPCC SRES and Special Report on Technology Transfer as well as IPCC Technical Paper I (1996) and Technical Paper IV (1997). Prof. Shukla has been a member of several prestigious National and International Policy Committees. He has led several international research projects and is a member of many international teams working on integrated assessment modeling and policy studies. He is a co-author/co-editor of fourteen books and numerous publications in international journals in the areas of energy, environment, climate change and development policies. He has guided over forty doctoral students. He obtained M.S (1976) and Ph.D. (1979) degrees from Stanford University, USA (Prof. Shukla can be contacted at: shukla@iima.ac.in).

Mr. Karthik Ganesan is a Research Fellow at CEEW. His research focus includes the development of long-term energy scenarios for India and energy efficiency improvements in the industrial sector in India. He also leads an independent effort to assess GHG emissions from the industrial sector and improving transparency in reporting. In addition, he supports on-going work in the areas of power sector performance and reform and energy access indicators for rural Indian households. He also carried out a first-of-a-kind evaluation of the impact of industrial policies on the RE sector in India.

His published works include Factors Affecting household satisfaction with electricity supply in India (Nature), Energising India—Towards a Resilient and Equitable Energy System (SAGE, book), Rethink India's Energy Strategy (Nature), Co-location opportunities for RE and agriculture in North-western India: Trade-offs and Synergies (Applied Energy), Valuation of health impact of air pollution from thermal power plants (ADB), Technical feasibility of metropolitan siting of nuclear power plants (NUS), Prospects for Carbon Capture and Storage in SE Asia (ADB).

Karthik has a Master in Public Policy from the Lee Kuan Yew School of Public Policy at NUS, Singapore. His prior educational training resulted in an M.Tech in Infrastructure Engineering and a B.Tech in Civil Engineering from IIT Madras (Mr. Karthik can be contacted at: karthik.ganesan@ceew.in).

Index

N. Janardhanan et al. (eds.), *Resurgence of Nuclear Power*,
https://doi.org/10.1007/978-981-10-5029-9

MIX
Papier aus verantwortungsvollen Quellen
Paper from responsible sources
FSC® C105338

If you have any concerns about our products,
you can contact us on
ProductSafety@springernature.com

In case Publisher is established outside the EU,
the EU authorized representative is:
Springer Nature Customer Service Center GmbH
Europaplatz 3, 69115 Heidelberg, Germany

Printed by Libri Plureos GmbH
in Hamburg, Germany